W9-CUJ-706

CONTENTS

The construction industry is vital to the success of our country. Countless people go to work every day to build America. They work hard, are extremely productive, and produce impressive projects. Most people are aware of the dangers faced in construction, which is generally classified as a high-hazard industry. Construction companies face significant challenges in terms of safety issues on a daily basis. One of the primary challenges is worker safety and health. How workers respond to hazards can determine the success or failure of any construction company.

The United States Occupational Safety and Health Administration (OSHA) enforces safety and health regulations throughout the country. During 2010, federal and state OSHA agencies conducted about 33,000 inspections of construction sites. As a result of those inspections, over 100,000 citations were written and penalties in excess of $93 million were issued to construction companies.

The OSHA penalties are staggering; however, they amount to less than what companies in the United States pay for work-related injuries and illnesses. It is estimated that American companies spend about $4 billion per week, or $208 billion annually, for those injuries and illnesses.

The individual worker also pays when safety measures fail. Millions of workers are injured annually and about 5,000 workers lose their lives at their place of employment every year. These fatalities, injuries, and illnesses can be prevented.

This *DeWALT Construction Safety and OSHA Handbook* has been designed to assist construction companies and their employees to maintain safe jobsites. The information provided here is based on the federal OSHA regulations: CFR Part 1926. OSHA regulations and resources can be found at http://www.osha.gov. Samples of the OSHA resources cited by the author can be found at the back of this book and on dewalt.com/guides. For those working under state-run OSHA plans, the rules and regulations may differ somewhat, but they must be at least as effective as the federal OSHA regulations.

CHAPTER 1

WORKPLACE SAFETY— TRAGEDY TO PROGRESS

Workplace safety has not always been what it is today. As the United States entered the Industrial Revolution, machines and other technological advances brought new dangers to the workplace. The government did not regulate and enforce employee safety and health the way it does now. This area was left up to business owners and industrialists. Regrettably, it took many years, thousands of disabling injuries, and many deaths before workplace safety became a priority.

THE TRIANGLE SHIRTWAIST FACTORY FIRE

In 1911, a factory fire in New York City changed the way our country viewed workplace safety. Three floors of a garment factory in lower Manhattan went up in flames after someone tossed out a match or cigarette. Unfortunately for the workers, there was little chance of escape. With only one fire escape, which collapsed during the rescue, and the building's many locked doors, 146 workers perished in the fire. Compounding the locked doors and collapsed fire escape were many large machines and huge tables that blocked the victims' path. As the workers scrambled to escape, they were crushed against the locked doors as they struggled to open them.

There was little the workers could do to fight the Triangle fire from the inside, and firefighters who responded to the fire did not have ladders long enough to reach the eighth, ninth, and tenth floors where the workers were trapped.

During this time in our history, companies were more concerned about theft and loss of productivity than preventing workplace hazards. At one time, construction companies planned on losing one worker for every million dollars spent on a project. Workers were often seen as disposable and losing them to injury or death usually didn't cost the company any money.

The Triangle fire and the outcome were preventable. The community's grief turned to a resolution for improvement. As the immigrant workers in New York City rallied against the working conditions that caused the fire, outcry from the rest of the public grew. Along with the U.S. Secretary of Labor, the governor of New York helped organize new workplace safety standards into law for the state of New York.

The Triangle Shirtwaist Factory fire brought national attention to workplace safety and health, but it was not the first or last large-scale workplace tragedy. This and many other tragedies brought changes to the workplace. Families and friends protested, lawsuits were filed, and many began to see that basic safety and health principles could have prevented the tragedies, or at the very least lessened their impact. Industry, government, workers, and communities recognized the need for a change.

PROGRESS

Workplace safety was a topic of conversation throughout the Industrial Revolution; however, the requirements were few and work-related accidents were relatively cheap for employers. With public outcry, progress began to take shape. Workers' compensation laws, the labor union movement, and, later, the creation of OSHA forced industrialists to become concerned about workplace safety and health.

Workers' Compensation

Work-related accidents began to cost companies more money with the passage of workers' compensation laws. The U.S. Congress passed in 1908 a federal employers' liability law. This law increased the cost of accidents tenfold. Then, in 1910, the state of New York became the first to pass a workers' compensation law. This new law automatically covered all injuries. Workers were no longer required to sue their employer in court after they were injured. The New York law was the first of many to come—between 1911 and 1921, 44 states passed workers' compensation laws. The current workers' compensation system is covered in Chapter 3.

Union Involvement

The labor union movement in the United States was underway well before the Triangle Shirtwaist Fire, but that single event galvanized workers to unionize and demand workplace rights. A key goal of the labor union movement was workplace safety and health. Labor unions were able to push safety to the forefront through collective bargaining. Safety and health regulations adopted as a result of the labor union movement were the basis for the federal regulations that were adopted in the 1970s.

OCCUPATIONAL SAFETY AND HEALTH ADMINISTRATION (OSHA)

As pressure from the public and the unions continued to increase, the U.S. government began looking into a system for creating and enforcing safety and health rules. Before 1970, employee safety and health programs relied on the states to enforce rules and regulations. The first state programs were in New York, California, Minnesota, Colorado, Washington, Georgia, and Nevada. The fact that the rules lacked comprehensiveness and uniformity hampered the state-run programs, resulting in a big push for nationalized safety and health regulation during the 1960s.

Congress debated and eventually, on December 29, 1970, President Nixon signed into law the Occupational Safety and Health Act. The OSH Act created the Occupational Safety and Health Administration (OSHA). The full text of the OSH Act can be found on the OSHA web page: http://www.osha.gov. The purpose of this act is stated as[1]:

To assure safe and healthful working conditions for working men and women; by authorizing enforcement of the standards developed under the Act; by assisting and encouraging the States in their efforts to assure safe and healthful working conditions; by providing for research, information, education, and training in the field of occupational safety and health; and for other purposes.

Over the years, OSHA would prove to have a huge impact on the safety and health of American workers. Through the creation of regulations and their enforcement, companies have been forced to keep their workers safe. This guide will take you through some of these regulations and provide you with a rational explanation of what construction companies need to do to protect their employees and comply with the law.

---

[1]Occupational Safety & Health Administration, U.S. Department of Labor. http://www.osha.gov.

One primary regulation stated in the OSH Act, commonly referred to as the "General Duty Clause," is as follows[2]:

5. Duties

    (a) Each employer —

        (1) shall furnish to each of his employees employment and a place of employment which are free from recognized hazards that are causing or are likely to cause death or serious physical harm to his employees;

        (2) shall comply with occupational safety and health standards promulgated under this Act.

    (b) Each employee shall comply with occupational safety and health standards and all rules, regulations, and orders issued pursuant to this Act which are applicable to his own actions and conduct.

For the purpose of the act, an employer is defined as[3]:

any "person engaged in a business affecting commerce who has employees, but does not include the United States or any State or political subdivision of a State."

The General Duty Clause can also be called the "Catch-All Standard." It provides OSHA the background to create safety and health regulations designed to protect employees from recognized hazards.

The next chapter will present an introduction to OSHA. Since the passage of the OSH Act in 1970 and the subsequent creation of OSHA, workers have had safer places to work in America. Unfortunately, about 5,000 workers still lose their lives annually. OSHA has focused on the construction industry, which it generally considers to be a high-hazard industry. On a daily basis, construction companies must be diligent to protect their workers.

---

[2]Occupational Safety & Health Administration, U.S. Department of Labor. *General Duty Clause*, Section 5(a)(1) OSH Act. http://www.osha.gov.

[3]Occupational Safety & Health Administration, U.S. Department of Labor. *OSH Act*, Section 3(5). http://www.osha.gov.

CHAPTER **2**

INTRODUCTION
TO OSHA

As discussed in the previous chapter, prior to 1970 there were no national safety and health regulations. The passage of the OSH Act and the creation of OSHA helped change that forever.

NEED FOR LEGISLATION

There was a lot of consternation in America when the OSH Act was passed. The need for this legislation was constantly questioned. At the time the OSH Act was passed, about 90 million Americans were in the workforce and these workers were considered our most valuable national resource. Prior to the passage of the Act, approximately 14,000 workers perished each and every year. Also, a staggering 2.5 million workers were disabled each year and 300,000 new occupational disease cases were reported annually. In the 40 years since the passage of the OSH Act, workers continued to die at work; however, the annual worker deaths have decreased by about 60%.

PURPOSE OF OSHA

OSHA has the following purposes:

- Reduction of workplace hazards
- Implementation of new or improve existing safety and health programs
- Providing for research to solve occupational safety and health problems
- Establishment of employer and employee responsibilities for safety and health conditions
- Building on employer and employee safety and health initiatives
- Focusing on occupational health to prevent diseases occurring in the work environment
- Establishing training programs to increase the number and competence of occupational safety and health personnel
- Developing mandatory job safety and health standards and enforcing them effectively
- Developing recordkeeping and reporting requirements
- Providing for the development, analysis, evaluation, and approval of state occupational safety and health programs

REGULATIONS FOR THE CONSTRUCTION INDUSTRY

Regulations for the construction industry are found in CFR Part 1926. The regulations are organized into major categories called subparts. The following are the subparts for CFR Part 1926:

Subpart A	General
Subpart B	General Interpretations
Subpart C	General Safety and Health Provisions
Subpart D	Occupational Health and Environmental Controls
Subpart E	Personal Protective and Life Saving Equipment
Subpart F	Fire Protection and Prevention
Subpart G	Signs, Signals, and Barricades
Subpart H	Materials Handling, Storage, Use and Disposal
Subpart I	Tools—Hand and Power
Subpart J	Welding and Cutting
Subpart K	Electrical
Subpart L	Scaffolds
Subpart M	Fall Protection
Subpart N	Cranes, Derricks, Hoists, Elevators, and Conveyors
Subpart O	Motor Vehicles, Mechanized Equipment, and Marine Operations
Subpart P	Excavations
Subpart Q	Concrete and Masonry Construction
Subpart R	Steel Erection
Subpart S	Tunnels and Shafts, Caissons, Cofferdams, and Compressed Air
Subpart U	Blasting and Use of Explosives
Subpart V	Power Transmission and Distribution
Subpart W	Rollover Protective Structures; Overhead Protection
Subpart X	Stairways and Ladders
Subpart Y	Commercial Driving Operations
Subpart Z	Toxic and Hazardous Substances
Subparts AA–BB	RESERVED
Subpart CC	Cranes and Derricks in Construction
Subpart DD	Cranes and Derricks Used in Demolition and Underground Construction

Each of these subparts contains the regulations related to their respective titles. In addition to regulations, OSHA has issued compliance directives and interpretation letters. These documents will assist construction companies in making sure they are in compliance with OSHA.

STATE OSHA PLANS

The big push for nationalized safety and health regulations resulted in the passage of the OSH Act. However, many states fought for the right to keep or implement their own safety and health programs. These states were successful and so the act contains the following text: [The Act] "encourages the States to assume the fullest responsibility for the administration and enforcement of their occupational safety and health laws." The OSH Act also provides grant money for states to identify needs and determine responsibilities for occupational safety and health. With this, many states kept their original plans and had only to modify their regulations.

The following states have approved state plans:

- Alaska
- Arizona
- California
- Connecticut
- Hawaii
- Illinois
- Indiana
- Iowa
- Kentucky
- Maryland
- Michigan
- Minnesota
- Nevada
- New Jersey
- New Mexico
- New York
- North Carolina
- Oregon
- Puerto Rico
- South Carolina
- Tennessee
- Utah
- Virgin Islands
- Virginia
- Vermont
- Washington
- Wyoming

NOTE

The Connecticut, Illinois, New Jersey, New York, and Virgin Islands plans cover public-sector (state and local government) employees only.

"At Least as Effective"

One key requirement for states retaining their own safety and health programs was that the state program must be at least as effective as the federal OSHA program. This does not mean that the regulations have to be identical; they just must be as effective. States also have the right to pass and enforce regulations stricter than the federal regulations found in CFR Part 1926. The majority of states have adopted the Federal regulations without change. However a few states, such as California, have adopted their own distinct regulations. This guide is based on the federal regulations. If you work in a state with its own plan, please check with that state to ensure compliance. State plans are subject to oversight by OSHA to determine if they are at least as effective. The following is a news release by OSHA concerning its oversight:

OSHA News Release—State Program Evaluation

Release Number: 10-1325-NAT
Sept. 28, 2010

US Labor Department's OSHA reports on state-run occupational safety and health programs—Agency calls for corrective actions to keep workers safe

WASHINGTON–The U.S. Department of Labor's Occupational Safety and Health Administration today announced that it has concluded a special evaluation of state-run occupational safety and health programs under its jurisdiction. Enhanced Federal Annual Monitoring and Evaluation reports provide detailed findings and recommendations on the operations of state-run OSHA programs in 25 states and territories. The enhanced review was initiated after a 2009 special OSHA report on Nevada's program, prompted by numerous construction-related fatalities in Las Vegas, identified serious operational deficiencies in that state.

"Our goal is to identify problems in state-run programs before they result in serious injuries or fatalities," said Assistant Secretary of Labor for OSHA Dr. David Michaels. "While we found many positives in the state programs, we also found deficiencies including concerns about identification of hazards, proper classification of violations, proposed penalty levels, and failure to follow up on violations to ensure that workplace safety and health problems are corrected."

The EFAME report and appendices for each of the 25 states, as well as each state's comment and fiscal year 2009 self-evaluation reports, are now available on OSHA's website at http://www.osha.gov.

States will have 30 days to provide a formal response, including a detailed corrective action plan for addressing findings and recommendations. Each state's formal response will be public information and available online as soon as it is received. The EFAME review also identified areas where states have adopted standards and procedures exceeding federal OSHA requirements, such as injury and illness prevention programs in California, Washington, Oregon, Minnesota, and other states; the adoption of a cranes and derricks rule prior to OSHA's in North Carolina, Washington, and Maryland; and Oregon's requirement that employers abate serious workplace violations during the contest period, a legal tool under consideration in Congress but still lacking in federal OSHA.

The review of the Hawaii program highlights significant performance problems resulting from staffing and funding cutbacks. OSHA is addressing these problems directly with the governor's office and has offered to provide supplemental federal enforcement assistance until the state can address its problems. If Hawaii is unable to present a reasonable strategy for expeditiously improving its worker safety and health oversight, consideration will be given to the state's current authority to operate its own program independently and could result in a federal takeover.

"We recognize that some of the problems we identified could stem from significant budget constraints in many of the states and may also be the result of less intensive federal oversight in recent years," Michaels added. "OSHA, through its regional offices, intends to provide assistance in the implementation of corrective actions and will work closely with state officials to review progress. We are confident that by working together to address identified problems, we can improve state operations and provide more consistent protection to all of America's workers."

The 25 states and territories evaluated are Alaska, Arizona, California, Connecticut, Hawaii, Indiana, Iowa, Kentucky, Maryland, Michigan, Minnesota, New Jersey, New Mexico, New York, North Carolina, Oregon, Puerto Rico, South Carolina, Tennessee, Utah, Vermont, U.S. Virgin Islands, Virginia, Washington, and Wyoming. No reports are being issued on the Nevada and Illinois state plans; a special study was issued on the Nevada state plan in October 2009, and the Illinois state plan was not approved until September 2009. The status of each state's efforts to improve its plans will be reflected in the fiscal year 2010 Federal Annual Monitoring and Evaluation report expected in 2011. For more information about those states operating their own plans, visit http://www.osha.gov/dcsp/osp/index.html.

When Congress enacted the Occupational Safety and Health Act of 1970, it created an opportunity for federal-state partnerships to promote safety and health. Section 18 of the law allows states to develop and enforce occupational safety and health standards in the context of an OSHA-approved state plan. Twenty-seven states and territories have sought and obtained approval. Twenty-one states and Puerto Rico have complete programs covering both the private sector and state and local governments. Four states and the U.S. Virgin Islands have programs that limit coverage to public-sector employees. Currently, state plans deliver the OSHA program to 40% of the nation's workplaces, and the federal program is responsible for the other 60%.

State plan standards and enforcement must be at least as effective as the federal OSHA plan in providing safe and healthful employment to workers. In addition, state plans operate under authority of state law, not delegated federal authority. Thus, in order to operate its own plan, a state must enact an equivalent of the federal OSH Act and must use administrative and regulatory procedures to adopt its own standards, regulations, and operating procedures, all of which must be updated within 6 months of any change in the federal program.

Under the Occupational Safety and Health Act of 1970, employers are responsible for providing safe and healthful workplaces for their employees. OSHA's role is to assure these conditions for America's working men and women by setting and enforcing standards, and providing training, education, and assistance. For more information, visit http://www.osha.gov.

ENFORCEMENT AND INSPECTIONS

The OSH Act granted authority to OSHA to enforce the safety and health regulations found in CFR Part 1926. OSHA uses highly trained and experienced compliance officers to inspect workplaces throughout the United States.

OSHA Inspection Priorities

Not all 111 million workplaces covered by the Act can be inspected immediately. The worst situations need attention first. OSHA, therefore, has established a system of inspection priorities. The following are the inspection priorities laid forth in OSHA Publication 2098[1]:

Imminent Danger	Imminent danger situations receive top priority. An imminent danger is any condition where there is reasonable certainty that a danger exists that can be expected to cause death or serious physical harm immediately or before the danger can be eliminated through normal enforcement procedures.
	If a compliance officer finds an imminent danger situation, he or she will ask the employer to voluntarily abate the hazard and remove endangered employees from exposure.
	Should the employer fail to do this, OSHA, through the regional solicitor, may apply to the Federal District Court for an injunction prohibiting further work as long as unsafe conditions exist.
Catastrophes and Fatal Accidents	Second priority goes to the investigation of fatalities and accidents resulting in a death or hospitalization of three or more employees. The employer must report such catastrophes to OSHA within 8 hours. OSHA investigates to determine the cause of these accidents and whether existing OSHA standards were violated.
Complaints and Referrals	Third priority goes to formal employee complaints of unsafe or unhealthful working conditions and to referrals from any source about a workplace hazard.

---

[1]Occupational Safety & Health Administration, U.S. Department of Labor. *OSHA Inspections*, Publication 2098, revised 2002. http://www.osha.gov.

	The OSH Act gives each employee the right to request an OSHA inspection when the employee believes he or she is in imminent danger from a hazard or when he or she thinks that there is a violation of an OSHA standard that threatens physical harm. OSHA will maintain confidentiality if requested, inform the employee of any action it takes regarding complaints, and, if requested, hold an informal review of any decision not to inspect.
Programmed Inspections	Next in priority are programmed inspections aimed at specific high-hazard industries, workplaces, occupations, or health substances, or other industries identified in OSHA's current inspection procedures. OSHA selects industries for inspection on the basis of factors such as the injury incidence rates, previous citation history, employee exposure to toxic substances, or random selection. OSHA also may develop special emphasis programs that are local, regional, or national in scope, depending on the distribution of the workplaces involved. OSHA normally will conduct comprehensive safety inspections in manufacturing in those establishments with lost-workday injury rates at or above the Bureau of Labor Statistics' (BLS) national rate for manufacturing currently in use by OSHA. States with their own plans may use somewhat different systems to identify industries for inspection.
Follow-up Inspections	A follow-up inspection determines if the employer has corrected previously cited violations. If an employer has failed to abate a violation, the compliance officer informs the employer that he or she is subject to "Failure to Abate" alleged violations. This involves proposed additional daily penalties until the employer corrects the violation.

Because the construction industry is classified as high hazard, construction companies are inspected more often than companies in other industries. Construction companies should be prepared to be inspected by OSHA. For information on what to do during an OSHA inspection, see Chapter 10.

OSHA INSPECTION PROCESS

OSHA has a right to inspect a jobsite without advance notice. The following is the general process OSHA follows when it inspects a jobsite, as described in OSHA Publication 2098[2]:

Arrival of CSHO (Compliance Safety and Health Officer)	When the OSHA compliance officer arrives at the jobsite, he or she displays official credentials and asks to meet an appropriate employer representative. Employers should always ask to see the compliance officer's credentials. Employers may verify the OSHA federal or state compliance officer credentials by calling the nearest federal or state OSHA office. Compliance officers may not collect a penalty at the time of the inspection or promote the sale of a product or service at any time; anyone who attempts to do so is impersonating a government inspector and the employer should contact the FBI or local law enforcement officials immediately.

---

[2]Occupational Safety & Health Administration, U.S. Department of Labor. *OSHA Inspections*, Publication 2098, revised 2002. http://www.osha.gov.

Opening Conference	In the opening conference, the compliance officer explains how the jobsite was selected and the likely scope of the inspection.
	The compliance officer explains the purpose of the visit, the scope of the inspection, and the standards that apply. The compliance officer gives the employer information on how to get a copy of applicable safety and health standards as well as a copy of any employee complaint that may be involved (with the employee's name deleted, if the employee requests anonymity).
	The compliance officer asks the employer to select an employer representative to accompany the compliance officer during the inspection.
	The compliance officer also gives an authorized employee representative the opportunity to attend the opening conference and accompany the compliance officer during the inspection. If a recognized bargaining agent represents the employees, the agent ordinarily will designate the employee representative to accompany the compliance officer. Similarly, if there is a safety committee, the employee members of that committee will designate the employee representative (in the absence of a recognized bargaining agent). Where neither employee group exists, the employees themselves may select an employee representative, or the compliance officer may determine if any employee suitably represents the interest of other employees.
	The OSH Act does not require an employee representative for each inspection. Where there is no authorized employee representative, however, the compliance officer must consult with a reasonable number of employees concerning safety and health matters in the workplace.
Inspection Tour	After the opening conference, the compliance officer and accompanying representatives proceed through the jobsite to inspect work areas for safety and health hazards.
	The compliance officer determines the route and duration of the inspection. While talking with employees, the compliance officer makes every effort to minimize any work interruptions. The compliance officer observes safety and health conditions and practices; consults with employees privately, if necessary; takes photos, records video, and instrument readings; examines records; collects air samples; measures noise levels; surveys existing engineering controls; and monitors employee exposure to toxic fumes, gases, and dusts.
	An inspection tour may cover part or all of a jobsite, even if the inspection resulted from a specific complaint, fatality, or catastrophe. If the compliance officer finds a violation in open view, he or she may ask permission to expand the inspection.
	The compliance officer keeps all trade secrets observed confidential.
	The compliance officer consults employees during the inspection tour. He or she may stop and question workers, in private, about safety and health conditions and practices in their workplaces. Each employee is protected under the OSH Act from discrimination and retaliation by the employer for exercising his or her safety and health rights.
	OSHA places special importance on posting and recordkeeping requirements. The compliance officer will inspect records of deaths, injuries, and illnesses that the employer is required to keep. He or she will check to see that a summary of the OSHA Form Number 300 [see Resource A located at the back

of this book, or on dewalt.com/guides] is posted as required and that the OSHA workplace poster [see Resource B located at the back of this book, or on dewalt.com/guides], which explains employees' safety and health rights, is prominently displayed. Where records of employee exposure to toxic substances and harmful physical agents are required, the compliance officer will examine them for compliance with the recordkeeping requirements. OSHA recordkeeping requirements are covered in more detail later in this chapter.

The compliance officer also requests a copy of the employer's Hazard Communication Program. Under OSHA's Hazard Communication Standard, employers must establish a written, comprehensive communication program that includes provisions for container labeling, material safety data sheets, and an employee training program. The program must contain a list of the hazardous chemicals in each work area and the means the employer will use to inform employees of the hazards associated with these chemicals. [More information on OSHA's Hazard Communication Standard can be found in Resource C located at the back of this book, or on dewalt.com/guides.]

During the course of the inspection, the compliance officer will point out to the employer any unsafe or unhealthful working conditions observed. At the same time, the compliance officer will discuss possible corrective action if the employer so desires.

Some apparent violations detected by the compliance officer can be corrected immediately. When the employer corrects them on the spot, the compliance officer records such corrections to help in judging the employer's good faith in compliance. Although corrected, the apparent violations will serve as the basis for a citation and, if appropriate, a notice of proposed penalty. OSHA may reduce the penalties for some types of violations if they are corrected immediately.

Closing Conference

At the conclusion of the inspection, the compliance officer conducts a closing conference with the employer, employees, and/or the employees' representative.

The compliance officer gives the employer and all other parties involved a copy of a publication titled "Employer Rights and Responsibilities Following an OSHA Inspection" for their review and discussion.

The compliance officer discusses with the employer all unsafe or unhealthful conditions observed during the inspection and indicates all apparent violations for which he or she may issue or recommend a citation and a proposed penalty. The compliance officer will not indicate any specific proposed penalties but will inform the employer of appeal rights.

During the closing conference, the employer may wish to produce records to show compliance efforts and provide information that can help OSHA determine how much time may be needed to abate an alleged violation.

When appropriate, the compliance officer may hold more than one closing conference. This is usually necessary when the inspection includes an evaluation of health hazards, after a review of additional laboratory reports, or after the compliance officer obtains additional factual evidence while concluding an accident investigation.

	If an employee representative does not participate in either the opening or the closing conference held with the employer, the compliance officer holds a separate discussion with the employee representative, if requested, to discuss matters of direct interest to employees.
After the Inspection	After the compliance officer reports their findings, the Area Director determines whether he or she will issue citations and/or propose penalties.

CITATIONS AND PENALTIES

Citations inform the employer and employees of the regulations and standards alleged to have been violated and of the proposed length of time allotted to correct the alleged hazards. The employer will receive citations and notices of proposed penalties by certified mail. The employer must post a copy of each citation at or near the place a violation occurred for 3 days or until the violation is abated, whichever is longer.

If your company receives a citation from OSHA, the citations will be categorized and penalized as described in Publication 2098[3]:

Other-Than-Serious Violation	A violation that has a direct relationship to job safety and health, but probably would not cause death or serious physical harm. OSHA may assess a penalty up to $7,000 for each violation. OSHA may adjust a penalty for an other-than-serious violation downward based on the employer's good faith, history of previous violations, and size of business.
Serious Violation	A violation where there is a substantial probability that death or serious physical harm could result. OSHA assesses the penalty for a serious violation up to $7,000 for each violation depending on the gravity of the violation. OSHA may adjust a penalty for a serious violation downward based on the employer's good faith, history of previous violations, and size of business.
Willful Violation	A violation that the employer intentionally and knowingly commits. The employer is aware that a hazardous condition exists, knows that the condition violates a standard or other obligation of the Act, and makes no reasonable effort to eliminate it. OSHA may propose penalties of up to $70,000 for each willful violation. The minimum willful penalty is $5,000. An employer who is convicted in a criminal proceeding of a willful violation of a standard that has resulted in the death of an employee may be fined up to $250,000 (or $500,000 if the employer is a corporation), imprisoned up to 6 months or both. A second conviction doubles the possible term of imprisonment.
Repeated Violation	A violation of any standard, regulation, rule, or order where, upon re-inspection, a substantially similar violation is found and the original citation has become a final order. Violations can bring a fine or up to $70,000 for each such violation. The repeat violation can be cited for up to 5 years after the original.
Failure-to-Abate	A failure to correct a prior violation may bring a civil penalty of up to $7,000 for each day that the violation continues beyond the prescribed abatement date.

---

[3]Occupational Safety & Health Administration, U.S. Department of Labor. *OSHA Inspections*, Publication 2098, revised 2002. http://www.osha.gov.

Additional violations for which OSHA may issue citations and proposed penalties are as follows:

- Falsifying records, reports, or applications can, upon conviction, bring a criminal fine of $10,000 or up to 6 months in jail, or both
- Violating posting requirements may bring a civil penalty of $7,000
- Assaulting a compliance officer or otherwise resisting, opposing, intimidating, or interfering with a compliance officer in the performance of his or her duties is a criminal offense and is subject to a fine of not more than $5,000 and imprisonment for not more than 3 years

Citations and penalty procedures may differ somewhat in states with their own plans.

Citation Appeal Process

Informal Conference

When issued a citation and notice of proposed penalty, an employer may request an informal conference with OSHA's area director to discuss the case. Attending the informal conference is strongly encouraged. The informal conference with the area director gives the employer the opportunity to discuss issues arising from the inspection that he or she wishes to discuss or to provide additional information. The Area Director is authorized to enter into settlement agreements that revise citations and penalties to avoid prolonged legal disputes and that result in speedier hazard abatement.

Petition for Modification of Abatement

After receiving a citation, the employer must correct the cited hazard by the abatement date unless he or she contests the citation or abatement date. Factors beyond the employer's control, however, may prevent the completion of corrections by that date. In such a situation, the employer who has made a good-faith effort to comply may file a petition to modify the abatement date.

The written petition must specify the steps taken to achieve compliance, the additional time needed to comply, the reasons additional time is needed, and interim steps taken to safeguard employees against the cited hazard during the intervening period. The employer must certify that he or she posted a copy of the petition in a conspicuous place at or near each place where a violation occurred and that the employee representative received a copy of the petition.

Notice of Contest

If the employer decides to contest the citation, the abatement period, or the proposed penalty, he or she has 15 working days from the time the citation and proposed penalty are received to notify the OSHA area director in writing. Failure to do so results in the citation and proposed penalty becoming a final order of the Occupational Safety and Health Review Commission (OSHRC) without further appeal. An orally expressed disagreement will not suffice. This written notification is called a "Notice of Contest."

Although there is no specific format for the Notice of Contest, it must clearly identify the employer's basis for filing—the citation, notice of proposed penalty, abatement period, or notification of failure to correct violations.

The employer must give a copy of the Notice of Contest to the employees' authorized representative. If any affected employees are not represented by a recognized bargaining agent, the employer must post a copy of the notice in a prominent location in the workplace or give it personally to each unrepresented employee.

Review Procedure

If the employer files a written Notice of Contest within the required 15 working days, the OSHA area director forwards the case to OSHRC. The commission is an independent agency not associated with OSHA or the Department of Labor. The commission assigns the case to an administrative law judge.

OSHRC may schedule a hearing at a public place near the employer's workplace. The employer and the employee have the right to participate in the hearing; the OSHRC does not require them to be represented by attorneys.

Once the administrative law judge has ruled, any party to the case may request a further review by OSHRC. Any of the three OSHRC commissioners also may, at his or her own motion, bring a case before the commission for review. Employers and other parties may appeal commission rulings to the appropriate U.S. Court of Appeals.

Appeals in States with Their Own Plans

States with their own plans have a state system for review and appeal of citations, penalties, and abatement periods. The procedures are generally similar to federal OSHA's, but a state review board or equivalent authority hears cases.

MULTI-EMPLOYER WORKSITE POLICY

OSHA Directive Number CPL 2-0.124 addresses worksites with multiple employers. This directive has major implications for the construction industry as it allows OSHA to cite general contractors for the unsafe work practices of their subcontractors. The following are excerpts from this OSHA directive[4]:

On multi-employer worksites (in all industry sectors), more than one employer may be citable for a hazardous condition that violates an OSHA standard. A two-step process must be followed in determining whether more than one employer is to be cited.

Step 1: The first step is to determine whether the employer is a creating, exposing, correcting, or controlling employer. The definitions below explain and give examples of each. Remember that an employer may have multiple roles. Once you determine the role of the employer, go to Step Two to determine if a citation is appropriate (NOTE: only exposing employers can be cited for General Duty Clause violations).

Step 2: If the employer falls into one of these categories, it has obligations with respect to OSHA requirements. Step Two is to determine if the employer's actions were sufficient to meet those obligations. The extent of the actions required of employers varies based on which category applies.

The Creating Employer

Step 1: Definition: The employer that caused a hazardous condition that violates an OSHA standard.

Step 2: Actions Taken: Employers must not create violative conditions. An employer that does so is citable even if the only employees exposed are those of other employers at the site.

Example 1: Employer Host operates a factory. It contracts with Company S to service machinery. Host fails to cover drums of a chemical despite S's repeated requests that it do so. This results in airborne levels of the chemical that exceed the Permissible Exposure Limit.

Analysis: Step 1: Host is a creating employer because it caused employees of S to be exposed to the air contaminant above the PEL. Step 2: Host failed to implement measures to prevent the accumulation of the air contaminant. It could have met its OSHA obligation by implementing the simple engineering control of covering the drums. Having failed to implement a feasible engineering control to meet the PEL, Host is citable for the hazard.

Example 2: Employer M hoists materials onto Floor 8, damaging perimeter guardrails. Neither its own employees nor employees of other employers are exposed to the hazard. It takes effective steps to keep all employees, including those of other employers, away from the unprotected edge and informs the controlling employer of the problem. Employer M lacks authority to fix the guardrails itself.

Analysis: Step 1: Employer M is a creating employer because it caused a hazardous condition by damaging the guardrails. Step 2: While it lacked the authority to fix the guardrails, it took immediate and effective steps to

---

[4]Occupational Safety & Health Administration, U.S. Department of Labor. *CPL 2-0.124, Multi-Employer Citation Policy.* http://www.osha.gov.

keep all employees away from the hazard and notified the controlling employer of the hazard. Employer M is not citable since it took effective measures to prevent employee exposure to the fall hazard.

The Exposing Employer

Step 1: Definition: An employer whose own employees are exposed to the hazard.

Step 2: Actions taken: If the exposing employer created the violation, it is citable for the violation as a creating employer. If the violation was created by another employer, the exposing employer is citable if it:

1. knew of the hazardous condition or failed to exercise reasonable diligence to discover the condition, and

2. failed to take steps consistent with its authority to protect is employees.

If the exposing employer has authority to correct the hazard, it must do so. If the exposing employer lacks the authority to correct the hazard, it is citable if it fails to do each of the following:

1. ask the creating and/or controlling employer to correct the hazard;

2. inform its employees of the hazard; and

3. take reasonable alternative protective measures.

In extreme circumstances (e.g., imminent danger situations), the exposing employer is citable for failing to remove its employees from the job to avoid the hazard.

Example 3: Employer Sub S is responsible for inspecting and cleaning a work area in Plant P around a large, permanent hole at the end of each day. An OSHA standard requires guardrails. There are no guardrails around the hole and Sub S employees do not use personal fall protection, although it would be feasible to do so. Sub S has no authority to install guardrails. However, it did ask Employer P, which operates the plant, to install them. P refused to install guardrails.

> **Analysis:** Step 1: Sub S is an exposing employer because its employees are exposed to the fall hazard. Step 2: While Sub S has no authority to install guardrails, it is required to comply with OSHA requirements to the extent feasible. It must take steps to protect its employees and ask the employer that controls the hazard—Employer P—to correct it. Although Sub S asked for guardrails, since the hazard was not corrected, Sub S was responsible for taking reasonable alternative protective steps, such as providing personal fall protection. Because that was not done, Sub S is citable for the violation.

Example 4: Unprotected rebar on either side of an access ramp presents an impalement hazard. Sub E, an electrical subcontractor, does not have the authority to cover the rebar. However, several times Sub E asked the general contractor, Employer GC, to cover the rebar. In the meantime, Sub E instructed its employees to use a different access route that avoided most of the uncovered rebar and required them to keep as far from the rebar as possible.

> **Analysis:** Step 1: Since Sub E employees were still exposed to some unprotected rebar, Sub E is an exposing employer. Step 2: Sub E made a good faith effort to get the general contractor to correct the hazard and took feasible measures within its control to protect its employees. Sub E is not citable for the rebar hazard.

The Correcting Employer

Step 1: Definition: An employer who is engaged in a common undertaking, on the same worksite, as the exposing employer and is responsible for correcting a hazard. This usually occurs where an employer is given the responsibility of installing and/or maintaining particular safety/health equipment or devices.

Step 2: Actions taken: The correcting employer must exercise reasonable care in preventing and discovering violations and meet its obligations of correcting the hazard.

Example 5: Employer C, a carpentry contractor, is hired to erect and maintain guardrails throughout a large, 15-story project. Work is proceeding on all floors. C inspects all floors in the morning and again in the afternoon each day. It also inspects areas where material is delivered to the perimeter once the material vendor is finished delivering material to that area. Other subcontractors are required to report damaged/missing guardrails to the general contractor, who forwards those reports to C. C repairs damaged guardrails immediately after finding them and immediately after they are reported. On this project few instances of damaged guardrails have occurred other than where material has been delivered. Shortly after the afternoon inspection of Floor 6, workers moving equipment accidentally damage a guardrail in one area. No one tells C of the damage and C has not seen it. An OSHA inspection occurs at the beginning of the next day, prior to the morning inspection of Floor 6. None of C's own employees are exposed to the hazard, but other employees are exposed.

> **Analysis:** Step 1: C is a correcting employer since it is responsible for erecting and maintaining fall protection equipment. Step 2: The steps C implemented to discover and correct damaged guardrails were reasonable in light of the amount of activity and size of the project. It exercised reasonable care in preventing and discovering violations; it is not citable for the damaged guardrail since it could not reasonably have known of the violation.

The Controlling Employer

Step 1: Definition: An employer who has general supervisory authority over the worksite, including the power to correct safety and health violations itself or require others to correct them. Control can be established by contract or, in the absence of explicit contractual provisions, by the exercise of control in practice. Descriptions and examples of different kinds of controlling employers are given below.

Step 2: Actions Taken: A controlling employer must exercise reasonable care to prevent and detect violations on the site. The extent of the measures that a controlling employer must implement to satisfy this duty of reasonable care is less than what is required of an employer with respect to protecting its own employees. This means that the controlling employer is not normally required to inspect for hazards as frequently or to have the same level of knowledge of the applicable standards or of trade expertise as the employer it has hired.

Factors Relating to Reasonable Care Standard. Factors that affect how frequently and closely a controlling employer must inspect to meet its standard of reasonable care include:

- The scale of the project;
- The nature and pace of the work, including the frequency with which the number or types of hazards change as the work progresses;
- How much the controlling employer knows both about the safety history and safety practices of the employer it controls and about that employer's level of expertise.
- More frequent inspections are normally needed if the controlling employer knows that the other employer has a history of non-compliance. Greater inspection frequency may also be needed, especially at the beginning of the project, if the controlling employer had never before worked with this other employer and does not know its compliance history.
- Less frequent inspections may be appropriate where the controlling employer sees strong indications that the other employer has implemented effective safety and health efforts. The most important indicator of an effective safety and health effort by the other employer is a consistently high level of compliance. Other indicators include the use of an effective, graduated system of enforcement for non-compliance with safety and health requirements coupled with regular jobsite safety meetings and safety training.

Evaluating Reasonable Care. In evaluating whether a controlling employer has exercised reasonable care in preventing and discovering violations, consider questions such as whether the controlling employer:

- Conducted periodic inspections of appropriate frequency;
- Implemented an effective system for promptly correcting hazards;
- Enforces the other employer's compliance with safety and health requirements with an effective, graduated system of enforcement and follow-up inspections.

Types of Controlling Employers

Control Established by Contract. In this case, the employer has a specific contract right to control safety: To be a controlling employer, the employer must itself be able to prevent or correct a violation or to require another employer to prevent or correct the violation. One source of this ability is explicit contract authority. This can take the form of a specific contract right to require another employer to adhere to safety and health requirements and to correct violations the controlling employer discovers.

Example 6: Employer GH contracts with Employer S to do sandblasting at GH's plant. Some of the work is regularly scheduled maintenance and so is general industry work; other parts of the project involve new work and are considered construction. Respiratory protection is required. Further, the contract explicitly requires S to comply with safety and health requirements. Under the contract GH has the right to take various actions against S for failing to meet contract requirements, including the right to have non-compliance corrected by using other workers and back-charging for that work. S is one of two employers under contract with GH at the worksite, where a total of five employees work. All work is done within an existing building. The number and types of hazards involved in S's work do not significantly change as the work progresses. Further, GH has worked with S over the course of several years. S provides periodic and other safety and health training and uses a graduated system of enforcement of safety and health rules. S has consistently had a high level of compliance at its previous jobs and at this site. GH monitors S by a combination of weekly inspections, telephone discussions and a weekly review of S's own inspection reports. GH has a system of graduated enforcement that it has applied to S for the few safety and health violations that had been committed by S in the past few years. Further, due to respirator equipment problems S violates respiratory protection requirements two days before GH's next scheduled inspection of S. The next day there is an OSHA inspection. There is no notation of the equipment problems in S's inspection reports to GH and S made no mention of it in its telephone discussions.

Analysis: Step 1: GH is a controlling employer because it has general supervisory authority over the worksite, including contractual authority to correct safety and health violations. Step 2: GH has taken reasonable steps to try to make sure that S meets safety and health requirements. Its inspection frequency is appropriate in light of the low number of workers at the site, lack of significant changes in the nature of the work and types of hazards involved, GH's knowledge of S's history of compliance and its effective safety and health efforts on this job. GH has exercised reasonable care and is not citable for this condition.

Example 7: Employer GC contracts with Employer P to do painting work. GC has the same contract authority over P as Employer GH had in Example 6. GC has never before worked with P. GC conducts inspections that are sufficiently frequent. Further, during a number of its inspections, GC finds that P has violated fall protection requirements. It points the violations out to P during each inspection but takes no further actions.

Analysis: Step 1: GC is a controlling employer since it has general supervisory authority over the site, including a contractual right of control over P. Step 2: GC took adequate steps to meet its obligation to discover violations. However, it failed to take reasonable steps to require P to correct hazards since it lacked a graduated system of enforcement. A citation to GC for the fall protection violations is appropriate.

Example 8: Employer GC contracts with Sub E, an electrical subcontractor. GC has full contract authority over Sub E, as in Example 6. Sub E installs an electric panel box exposed to the weather and implements an assured equipment grounding conductor program, as required under the contract. It fails to connect a grounding wire inside the box to one of the outlets. This incomplete ground is not apparent from a visual inspection. Further, GC inspects the site with a frequency appropriate for the site. It saw the panel box but did not test the outlets to determine if they were all grounded because Sub E represents that it is doing all of the required tests on all receptacles. GC knows that Sub E has implemented an effective safety and health program. From previous experience it also knows Sub E is familiar with the applicable safety requirements and is technically competent. GC had asked Sub E if the electrical equipment is OK for use and was assured that it is.

Analysis: Step 1: GC is a controlling employer since it has general supervisory authority over the site, including a contractual right of control over Sub E. Step 2: GC exercised reasonable care. It had determined that Sub E had technical expertise, safety knowledge and had implemented safe work practices. It conducted inspections with appropriate frequency. It also made some basic inquiries into the safety of the electrical equipment. Under these circumstances GC was not obligated to test the outlets itself to determine if they were all grounded. It is not citable for the grounding violation.

Control Established by a Combination of Other Contract Rights: Where there is no explicit contract provision granting the right to control safety, or where the contract says the employer does not have such a right, an employer may still be a controlling employer. The ability of an employer to control safety in this circumstance can result from a combination of contractual rights that, together, give it broad responsibility at the site involving almost all aspects of the job. Its responsibility is broad enough so that its contractual authority necessarily involves safety. The authority to resolve disputes between subcontractors, set schedules and determine construction sequencing are particularly significant because they are likely to affect safety.

Example 9: Construction manager M is contractually obligated to: set schedules and construction sequencing, require subcontractors to meet contract specifications, negotiate with trades, resolve disputes between subcontractors, direct work and make purchasing decisions, which affect safety. However, the contract states that M does not have a right to require compliance with safety and health requirements. Further, Subcontractor S asks M to alter the schedule so that S would not have to start work until Subcontractor G has completed installing guardrails. M is contractually responsible for deciding whether to approve S's request.

> **Analysis:** Step 1: Even though its contract states that M does not have authority over safety, the combination of rights actually given in the contract provides broad responsibility over the site and results in the ability of M to direct actions that necessarily affect safety. For example, M's contractual obligation to determine whether to approve S's request to alter the schedule has direct safety implications. M's decision relates directly to whether S's employees will be protected from a fall hazard. M is a controlling employer. Step 2: In this example, if M refused to alter the schedule, it would be citable for the fall hazard violation.

Example 10: Employer ML's contractual authority is limited to reporting on subcontractors' contract compliance to owner/developer O and making contract payments. Although it reports on the extent to which the subcontractors are complying with safety and health infractions to O, ML does not exercise any control over safety at the site.

> **Analysis:** Step 1: ML is not a controlling employer because these contractual rights are insufficient to confer control over the subcontractors and ML did not exercise control over safety. Reporting safety and health infractions to another entity does not, by itself (or in combination with these very limited contract rights), constitute an exercise of control over safety. Step 2: Since it is not a controlling employer it had no duty under the OSH Act to exercise reasonable care with respect to enforcing the subcontractors' compliance with safety; there is therefore no need to go to Step 2.
>
> Architects and Engineers: Architects, engineers, and other entities are controlling employers only if the breadth of their involvement in a construction project is sufficient to bring them within the parameters discussed above.

Example 11: Architect A contracts with owner O to prepare contract drawings and specifications, inspect the work, report to O on contract compliance, and to certify completion of work. A has no authority or means to enforce compliance, no authority to approve/reject work and does not exercise any other authority at the site, although it does call the general contractor's attention to observed hazards noted during its inspections.

> **Analysis:** Step 1: A's responsibilities are very limited in light of the numerous other administrative responsibilities necessary to complete the project. It is little more than a supplier of architectural services and conduit of information to O. Its responsibilities are insufficient to confer control over the subcontractors and it did not exercise control over safety. The responsibilities it does have are insufficient to make it a controlling employer. Merely pointing out safety violations did not make it a controlling employer. NOTE: In a circumstance such as this it is likely that broad control over the project rests with another entity. Step 2: Since A is not a controlling employer it had no duty under the OSH Act to exercise reasonable care with respect to enforcing the subcontractors' compliance with safety; there is therefore no need to go to Step 2.

Example 12: Engineering firm E has the same contract authority and functions as in Example 9.

> **Analysis:** Step 1: Under the facts in Example 9, E would be considered a controlling employer. Step 2: The same type of analysis described in Example 9 for Step 2 would apply here to determine if E should be cited.
>
> Control Without Explicit Contractual Authority. Even where an employer has no explicit contract rights with respect to safety, an employer can still be a controlling employer if, in actual practice, it exercises broad control over subcontractors at the site (see Example 9). NOTE: Citations should only be issued in this type of case after consulting with the Regional Solicitor's office.

Example 13: Construction manager MM does not have explicit contractual authority to require subcontractors to comply with safety requirements, nor does it explicitly have broad contractual authority at the site. However, it exercises control over most aspects of the subcontractors' work anyway, including aspects that relate to safety.

Analysis: Step 1: MM would be considered a controlling employer since it exercises control over most aspects of the subcontractor's work, including safety aspects. Step 2: The same type of analysis on reasonable care described earlier would apply to determine if a citation should be issued to this type of controlling employer.

OSHA REPORTING AND RECORDKEEPING REQUIREMENTS

Construction companies are required to report and record jobsite injuries and illnesses to OSHA. The following is an explanation of those requirements.

Jobsite Fatalities and Catastrophes

From 2005 through 2009, there was an average of over 5,300 work-related deaths in the United States. OSHA takes each and every one of these fatalities seriously and investigates them. OSHA requires all jobsite fatalities and catastrophes to be reported by the employer. The OSHA Regulation 29 CFR 1904.39(a) states[5]:

Within eight (8) hours after the death of any employee from a work-related incident or the in-patient hospitalization of three or more employees as a result of a work-related incident, you must orally report the fatality/multiple hospitalization by telephone or in person to the Area Office of the Occupational Safety and Health Administration (OSHA), U.S. Department of Labor, that is nearest to the site of the incident. You may also use the OSHA toll-free central telephone number, 1-800-321-OSHA (1-800-321-6742).

As presented earlier in this chapter, OSHA makes these investigations a high priority and will begin its investigation very soon after notification.

Recording Jobsite Injuries and Illnesses

Injuries and illnesses that occur at the workplace need to be recorded on the appropriate OSHA recordkeeping forms. (See Resource A located at the back of this book, or on dewalt.com/guides.) Over the past couple of years, OSHA has instituted an emphasis program on recordkeeping to ensure employers are complying with regulations. OSHA provides employers with the following information on their recordkeeping requirements, which are excerpts from OSHA Publication 3169 that describes the new recordkeeping requirements[6]:

Employers now have a new system for tracking workplace injuries and illnesses. OSHA's new recordkeeping log is easier to understand and to use. Written in plain language using a question and answer format, the revised recordkeeping rule answers questions about recording occupational injuries and illnesses and explains how to classify particular cases. Flowcharts and checklists make it easier to follow the recordkeeping requirements.

What Has Changed?

The new rule:

- Offers flexibility by letting employers computerize injury and illness records
- Updates three recordkeeping forms:
    - OSHA Form 300 (Log of Work-Related Injuries and Illnesses); simplified and reformatted to fit legal size paper
    - OSHA Form 301 (Injury and Illness Incident Report); includes more data about how the injury or illness occurred

---

[5]Occupational Safety & Health Administration, U.S. Department of Labor. *OSHA Regulation* 29 CFR 1904.39(a). http://www.osha.gov.
[6]Occupational Safety & Health Administration, U.S. Department of Labor. *OSHA Publication* 3169. http://www.osha.gov.

- OSHA Form 300A (Summary of Work-Related Injuries and Illnesses); separate form created to make it easier to calculate incidence rates
- Continues to exempt smaller employers (employers with 10 or fewer employees) from most requirements
- Changes the exemptions for employers in service and retail industries
- Clarifies the definition of work relationship, limiting the recording of pre-existing cases and adding new exceptions for some categories of injury and illness
- Includes new definitions of medical treatment, first aid, and restricted work to simplify recording decisions
- Eliminates different criteria for recording work-related injuries and work-related illnesses; one set of criteria will be used for both
- Changes the recording of needlestick injuries and tuberculosis
- Simplifies the counting of days away from work, restricted days, and job transfer
- Improves employee involvement and provides employees and their representatives with access to the information
- Protects privacy for injured and ill workers

Simplified, clearer definitions also make it easier for employers to determine which cases must be recorded. Posting an annual summary of workplace injuries and illnesses for a longer period of time improves employee access to information, and as employees learn how to report workplace injuries and illnesses, their involvement and participation increase.

Which Recordkeeping Requirements Apply to Me?

Reporting fatalities and catastrophes: All employers covered by the Occupational Safety and Health Act of 1970 must report to OSHA any workplace incident resulting in a fatality or the in-patient hospitalization of three or more employees within 8 hours.

Keeping injury and illness records: If you had 10 or fewer employees during all of the last calendar year or your business is classified in a specific low-hazard retail, service, finance, insurance, or real estate industry, you do not have to keep injury and illness records unless the Bureau of Labor Statistics or OSHA informs you in writing that you must do so.

Am I Exempt?

The construction industry is classified as a high hazard industry and therefore is not exempt from this rule. Employers with more than 10 employees must keep records.

What Do I Have to Do if I Am Not Exempt?

Employers not exempt from OSHA's recordkeeping requirements must prepare and maintain records of work-related injuries and illnesses. You need to review Title 29 of the Code of Federal Regulations (CFR) Part 1904, "Recording and Reporting Occupational Injuries and Illnesses," to see exactly which cases to record.

- Use the Log of Work-Related Injuries and Illnesses (Form 300) to list injuries and illnesses and track days away from work, restricted, or transferred
- Use the Injury and Illness Report (Form 301) to record supplementary information about recordable cases. You can use a workers' compensation or insurance form, if it contains the same information
- Use the Summary (Form 300A) to show totals for the year in each category. The summary is posted from February 1 to April 30 of each year

Each of these forms can be found in Resource A at the back of this book, or on dewalt.com/guides.

What Is So Important About Recordkeeping?

Recordkeeping is a critical part of an employer's safety and health efforts for several reasons:

- Keeping track of work-related injuries and illnesses can help you prevent them in the future
- Using injury and illness data helps identify problem areas. The more you know, the better you can identify and correct hazardous workplace conditions
- You can better administer company safety and health programs with accurate records
- As employee awareness about injuries, illnesses, and hazards in the workplace improves, workers are more likely to follow safe work practices and report workplace hazards. OSHA compliance officers can rely on the data to help them properly identify and focus on injuries and illnesses in a particular area. The agency also asks about 80,000 establishments each year to report the data directly to OSHA, which uses the information as part of its site-specific inspection targeting program. The Bureau of Labor Statistics (BLS) also uses injury and illness records as the source data for the Annual Survey of Occupational Injuries and Illnesses that shows safety and health trends nationwide and industrywide.

How Can I Get More Information on Recordkeeping?

The full preamble and text of the new rule is available online. You can find it by searching the Index on OSHA's website at http://www.osha.gov. You can also receive a copy of the regulation from OSHA's Office of Publications, P.O. Box 37535, Washington, DC 20013-7535; phone (202) 693-1888. If your workplace is in a state operating under an OSHA-approved plan, state plan recordkeeping regulations, although similar to federal ones, may have some more stringent or supplemental requirements such as reporting fatalities and catastrophes. Industry exemptions may also differ. Make sure you check your state OSHA plan.

Even with OSHA's inspection priorities, enforcement practices, and recordkeeping requirements, many of the 111 million U.S. workplaces covered by OSHA have never had any contact with OSHA. Business ethics have kept many companies focused on the safety of their workers. The threat of OSHA enforcement is also a powerful force, but the cost of insurance should not be overlooked. In the next chapter, the workers' compensation system will be presented. Chapter 3 will review how safety pays.

WORKERS' COMPENSATION INSURANCE

A driving force behind the improvement of workplace safety has been the insurance industry. Construction companies will generally carry multiple types of insurance to protect their interests. Of these various types of insurance programs, it is workers' compensation that has significantly improved workplace safety. Insurance is unquestionably a cost for construction companies; however, following proper safe work practices and complying with OSHA regulations can give construction companies a competitive advantage.

WORKERS' COMPENSATION OVERVIEW

Workers' compensation insurance laws are in place throughout the United States, either through federal or state laws. These laws were passed to assist injured or ill workers and make sure they receive appropriate medical care. The laws also give workers the right to receive monetary awards based upon the injury and any potential permanent disability. Prior to the passing of these laws, workers had to battle their employers in court for these rights. The laws also make available benefits for dependents of workers who are killed because of work-related accidents or illnesses. Several of these laws also shield employers and fellow workers by limiting the amount an injured employee can recover from an employer and by eliminating the liability of co-workers in most accidents.

Workers' compensation laws were passed to help employers and employees avoid long legal proceedings when a worker was injured or became ill due to a workplace safety issue. The system covered all work-related injuries and illnesses regardless of who was at fault. In exchange for this, the workers gave up their right to take their employer to court. Workers accepted this compromise because in many cases it was difficult to prove fault by the employer and the worker often lost the lawsuit, without even getting coverage for their medical expenses. Although some workers did win a large settlement from their employer, these successful cases were few and far between.

Under the workers' compensation system, workers can sue only if their injury was caused by someone other than their employer, as in the following examples.

- A worker is injured on a jobsite because of a separate contractor's unsafe practices, such as a roofing contractor dropping debris from the roof and striking an electrical contractor's employee in the arm, causing an injury. The injured worker would automatically be covered by the employer's workers' compensation insurance, but could potentially sue the roofing contractor.

- A worker is injured because of a defect or flaw in the equipment or machinery being used, such as a nail gun, though properly maintained, malfunctioning and striking a framing contractor's employee in the leg, causing an injury. The injured worker would automatically be covered by the employer's workers' compensation insurance, but could potentially sue the manufacturer of the nail gun.

Federal and State Laws

Both federal and state statutes apply to workers' compensation. The statutes are used as the foundation for regulations with which construction companies are required to comply. The federal statues cover federal employees and employees in a few distinct occupations or industries.

The majority of the states have enacted comprehensive state compensation programs that are applicable to most employers. These programs require employers to obtain workers' compensation insurance. This insurance is available through private insurance companies or, in some cases, through the state. Since each state has its own laws and manages its own workers' compensation system, the requirements for construction companies vary. The differences among states can be quite substantial. The information presented in this chapter is a broad overview of what most states offer; however, construction companies should consult with the state in which they are working to gather more specific information.

Workers' Compensation Coverage

The coverage provided by workers' compensation is for workers who are injured on the job, suffer a job-related illness, contract a disease because of a job-related exposure, or die from a job-related injury or illness. Workers are entitled to appropriate medical care, including rehabilitation. If the injury or illness results in a death, benefits will be available to the worker's family, based on a state-approved formula.

In the majority of states, medical expenses are completely covered, and workers' compensation insurance will also cover therapy as prescribed by a doctor. The therapy will vary based on the injury but can include physical therapy, massage therapy, and occupational rehabilitation.

Medical care is supervised and managed by an occupational health doctor. During care for some injuries, the doctor may determine that the worker has reached maximum medical improvement. If the maximum medical improvement leaves the worker with a permanent disability, other benefits may be paid to the worker. These disability benefits are usually determined by one of two methods:

- **Scheduled:** This method is used for the most common occupational injuries and illnesses, such as loss of limb, eyesight, or hearing. These injuries and illnesses are listed on a schedule that specifies the amount of benefits due to the worker.
- **Nonscheduled:** This method is used for other injuries and illnesses not included on the scheduled list. The amount of disability benefits will be based on a percentage of total disability as determined by the occupational health doctor.

Ineligible Workers

Each state defines which workers are eligible for workers' compensation. The majority of workers are covered; some, however, are ineligible. The following workers are sometimes ineligible:

- Sole proprietors
- Partners
- Independent contractors
- Real estate agents
- Part-time domestic employees in private homes
- Some farm workers
- Newspaper carriers
- Casual workers: workers whose work is not continuous or recurrent or whose compensation is below a certain monetary threshold

- Railroad employees: covered by federal statute
- Unpaid volunteers: workers who are not considered employees

Construction companies should check with their state agency to verify who are ineligible employees.

Noncovered Injuries

The workers' compensation system was designed to cover most injuries and illnesses that occur on the job. Workers are covered automatically and proof of negligence by the employer is not required. However, there are some exceptions. In these cases, the workers' compensation benefits may be denied or reduced:

- Injuries when workers are under the influence of drugs or alcohol
- Injuries that are self-inflicted, including those caused by a person fighting at work
- Injuries caused when workers are committing a serious crime
- Injuries suffered by workers when not on the job
- Injuries caused when workers violate a company policy

Days Away from Work

If the injury to a covered worker causes the worker to miss work, the worker will generally be eligible for some salary while off work. Workers are generally not entitled to the entire amount of their salary. Most of the time, they will be eligible to receive two-thirds of their normal salary. Many states have set a maximum weekly pay for workers missing days because of a job-related injury or illness. It therefore benefits workers to get back to work as soon as practical. This allows them to be a part of the company again and receive their full pay. Getting workers back to work also is very beneficial to the company. In addition to gaining increased production, companies can reduce their insurance liability by getting their workers back to work.

Modified Duty

One option for the injured worker, who is not completely healed, is modified duty. Modified duty allows an injured worker to return to work or remain at work performing physically appropriate modified duties according to their current capabilities as determined by the occupational health doctor. Construction companies should consult with the occupational health doctor or specialist when considering modified duty. Items the company should consider are:

- How the worker's current capabilities compare to the job requirement
- To what extent the job can be modified to meet the worker's capabilities
- What other modified-duty opportunities are there on a limited or full-time basis if modifying the worker's original job is not possible

Job modification is the removal or alteration of physical barriers that may prohibit an injured worker from performing the essential functions of the job. Successful job modifications are the result of a cooperative effort involving the following people:

- Rehabilitation specialist
- Employer
- Injured worker
- Occupational health doctor
- Other professionals as appropriate

Job modifications are developed and implemented keeping in mind the injured worker's limitations, restrictions, functional capacity, and physical capabilities. Many states have specific requirements and/or guidelines for modified duty. Companies should check with the state in which they are doing business for complete details.

SAFETY PAYS

It was mentioned earlier in this chapter that safety can give construction companies a competitive advantage. A safe workplace will reduce injuries and their overall impact on the company. When looking at the cost of workplace accidents, companies should consider both the direct and indirect costs of the accident.

- **Direct Cost:** Out-of-pocket expenses not included under the normal cost of doing business
- **Indirect Cost:** Costs not directly related to the accident, but that occur as a result of the accident

Here are some examples of direct and indirect costs:

Direct Costs	Indirect Costs
Medical costs	Clean-up time/days missed
Equipment damage	Legal fees
Ambulance service	Schedule delays
	Increased insurance premiums
	Training of new employees
	Accident investigations
	Unhappy clients

The usual estimate of the ratio between indirect and direct costs is between 4:1 and 15:1. The indirect cost is always the much higher cost and often difficult to fully capture.

Maintaining workplace safety is a sure way for a company to reduce the rate of incidents and accidents, thus reducing costs. Reduction in incidents and accidents will also save a construction company costs related to workers' compensation premiums. This can be accomplished if the company has been in business for a while and has a good safety record.

Experience Modification Rate (EMR)

Workers' compensation insurance providers use a formula provided by the state to determine what their clients will pay in premiums. The formulas are based upon the type of work the construction company performs, the wages they pay, and how much work the company does. Having a good safety record will allow a construction company to pay less.

The EMR is calculated based on the company's safety record over the previous three years. Claims and costs are analyzed to determine the safety record. Based upon that analysis, companies will be issued an EMR. An EMR of 1.0 is an indication that a construction company has an average safety record compared to other companies in its industry or classification. An EMR below 1.0 means that a construction company has a better safety record than other companies, and an EMR higher than 1.0 means the company has a poor safety record. The following table shows two identical construction companies with the same wages and insurance rates. The difference between them is their EMRs. As you can see, a company's safety record can become a competitive advantage or disadvantage. The construction company with a good safety record will save $225,000 just in insurance premiums over its competitor with a poor safety record. Will this reduction in costs help a company bid projects? Will it give a competitive advantage? Construction companies who focus on safety will realize that safety does indeed pay.

Type of Safety Record	GOOD	POOR
Wages	$3,000,000	$3,000,000
Insurance Rate	25%	25%
Insurance Costs	$750,000	$750,000
EMR	0.85	1.15
Insurance Premium	$637,500	$862,500
	=$225,000 in Savings	

Cost Containment

Construction companies working in certain states also have an additional option for reducing their workers' compensation premiums. Some state agencies have programs in place that allow a reduction in premiums based on a company's safety program and safety record. One of these programs is referred to as "cost containment." Here are some requirements for obtaining a cost containment certificate for companies working in Colorado:

- Formal declaration of a company-wide loss prevention and loss control policy, including designation of a representative responsible for overseeing the policy
- Formal creation of a documented safety committee and/or coordinator
- Clearly defined and conspicuously posted loss prevention rules
- All employees undergo initial and ongoing job/task training
- A cost containment program in cooperation with a designated medical provider who has knowledge of work injuries, fee schedules, and employer's operations is decisive on maximum-medical-improvement determinations, and communicates with the employer
- Written policies and procedures on claims management that include accident investigation and an early return to work/modified duty program
- A copy of the insurance carrier's loss reports covering the last three full policy years and the current policy year-to-date, together with the total number of hours worked by all employees during each of those policy years

As you now understand, safety can help companies save money and give them a competitive advantage. In the next few chapters, you will learn how to create a safety program, what rules apply to you, and how to manage safety on your sites. With this information, your safety record can improve dramatically.

CHAPTER 4

SAFETY AND HEALTH PROGRAMS

An important step for construction companies to take in order to improve their safety record is to create a safety and health program. These programs guide the safety efforts of all workers and management and should be implemented by all companies.

A safety and health program is a written document designed to detect and prevent workplace hazards. According to OSHA[1]:

An effective safety and health program depends on the credibility of management's involvement in the program; inclusion of employees in safety and health decisions; rigorous worksite analysis to identify hazards and potential hazards, including those which could result from a change in worksite conditions or practices; stringent prevention and control measures; and thorough training.

With that in mind, employers should develop a written safety and health program specific to their organization. For example, plumbing contractors should focus on the hazards and risks associated with their work, not the hazards and risks associated with steel erection. General contractors may need to develop safety and health programs that include a wide range of hazards and risks because of OSHA's Multi-Employer Worksite Policy (presented in Chapter 2).

OSHA GUIDANCE FOR DEVELOPING SAFETY AND HEALTH PROGRAMS

While developing a safety and health program, there are many sources of information for assistance. OSHA has developed guidelines for developing safety and health programs. These guidelines are currently voluntary; however OSHA is considering rule-making which may change these guidelines to mandatory. The primary elements of a safety and health program, according to OSHA, are:

- Management commitment and employee involvement
- Worksite analysis
- Hazard prevention and control
- Safety and health training

These guidelines and other information on developing safety and health programs can be found on the OSHA website at: http://www.osha.gov.

---

[1]Occupational Safety & Health Administration, U.S. Department of Labor. http://www.osha.gov.

CAL/OSHA GUIDANCE FOR DEVELOPING SAFETY AND HEALTH PROGRAMS

Another high-quality source for safety and health programs is California's OSHA program, Cal/OSHA. Chapter 2 presented information on state-run OSHA plans. Cal/OSHA is a state-run OSHA plan that enforces safety and health regulations in California. An Injury and Illness Prevention Program (IIPP) is another term for a safety and health program. In California, an IIPP is required by state law. Although these guidelines were developed for employers in the state of California, they are applicable to employers throughout the country. The following are some excerpts from Cal/OSHA's *Guide to Developing Your Workplace Injury and Illness Prevention Program*.[2]

Why Have a Workplace Injury and Illness Prevention Program?

Taking risks is a part of running a business, particularly for small business owners. You take risks in product development, marketing, and advertising in order to stay competitive. Some risks are just not worth the gamble. One of these is risking the safety and health of those who work for you.

Accidents Cost Money

Safety organizations, states, small business owners and major corporations alike now realize that the actual cost of a lost workday injury is substantial. For every dollar you spend on the direct costs of a worker's injury or illness, you will spend much more to cover the indirect and hidden costs. Consider what one lost workday injury would cost you in terms of:

- Productive time lost by an injured employee;
- Productive time lost by employees and supervisors attending the accident victim;
- Clean up and start up of operations interrupted by the accident;
- Time to hire or to retrain other individuals to replace the injured worker until his/her return;
- Time and cost for repair or replacement of any damaged equipment or materials;
- Cost of continuing all or part of the employee's wages, in addition to compensation;
- Reduced morale among your employees, and perhaps lower efficiency; increased workers' compensation insurance rates; and
- Cost of completing paperwork generated by the incident.

Controlling Losses

If you would like to reduce the costs and risks associated with workplace injuries and illnesses, you need to address safety and health right along with production.

Setting up an Injury and Illness Prevention Program helps you do this. In developing the program, you identify what has to be done to promote the safety and health of your employees and worksite, and you outline policies and procedures to achieve your safety and health goals.

What Is an Injury and Illness Prevention Program?

Your Injury and Illness Prevention Program must be a written plan that includes procedures and is put into practice. These elements are required:

- Management commitment/assignment of responsibilities;
- Safety communications system with employees;
- System for assuring employee compliance with safe work practices;

---

[2]Cal/OSHA. *Guide to Developing Your Workplace Injury and Illness Prevention Program with Checklists for Self-Inspection*, www.dir.ca.gov/dosh/.

- Scheduled inspections/evaluation system;
- Accident investigation;
- Procedures for correcting unsafe/unhealthy conditions;
- Safety and health training and instruction; and
- Recordkeeping and documentation.

Management Commitment/Assignment of Responsibilities

Your commitment to safety and health shows in every decision you make and every action you take. Your employees will respond to that commitment.

The person or persons with the authority and responsibility for your safety and health program must be identified and given management's full support. You can demonstrate your commitment through your personal concern for employee safety and health and by the priority you place on these issues.

If you want maximum production and quality, you need to control potential workplace hazards and correct hazardous conditions or practices as they occur or are recognized.

You must commit yourself and your company by building an effective Injury and Illness Prevention Program and integrating it into your entire operation.

This commitment must be backed by strong organizational policies, procedures, incentives, and disciplinary actions as necessary to ensure employee compliance with safe and healthful work practices.

They should include:

1. Establishment of workplace objectives for accident and illness prevention, like those you establish for other business functions such as sales or production for example: "Ten percent fewer injuries next year," "Reduce down-time due to poorly maintained equipment."

2. Emphasis on your staff's safety and health responsibilities and recognition by your supervisors and employees that they are accountable. Advise your management staff that they will be held accountable for the safety record of the employees working under them, and then back it up with firm action.

3. A means for encouraging employees to report unsafe conditions with assurance that management will take action.

4. Allocation of company financial, material, and personnel resources for:
   - Identifying and controlling hazards in new and existing operations and processes, and potential hazards.
   - Installing engineering controls.
   - Purchasing personal protective equipment.
   - Promoting and training employees in safety and health.

5. Setting a good example! If, for instance, you require hard hats to be worn in a specific area, then you and other management wear a hard hat in that area. If you and your management team do not support and participate in the program, you are doomed to failure from the start. It is especially important for plant supervisors and field superintendents to set a good example.

Safety Communications

Your program must include a system for communicating with employees—in a form readily understandable by all affected employees—on matters relating to occupational safety and health, including provisions designed to encourage employees to inform the employer of hazards at the worksite without fear of reprisal.

While this section does not require employers to establish labor-management safety and health committees, it is an option you should consider. If you choose to do so, remember that employers who elect to use a labor-management safety and health committee to comply with the communication requirements are presumed to be in substantial compliance if the committee:

1. Meets regularly but not less than quarterly.

2. Prepares and makes available to affected employees written records of the safety and health issues discussed at the committee meetings, and maintained for review by the Division upon request.

3. Reviews results of the periodic scheduled worksite inspections.

4. Reviews investigations of occupational accidents and causes of incidents resulting in occupational injury, occupational illness or exposure to hazardous substances, and where appropriate, submits suggestions to management for the prevention of future incidents.

5. Reviews investigations of alleged hazardous conditions brought to the attention of any committee member. When determined necessary by the committee, it may conduct its own inspection and investigation to assist in remedial solutions.

6. Submits recommendations to assist in the evaluation of employee safety suggestions.

7. Upon request of the Division, verifies abatement action taken by the employer to abate citations issued by the Division.

If your employees are not represented by an agreement with an organized labor union, and part of your employee population is unionized, the establishment of labor-management committees is considerably more complicated. You should request clarification from the Cal/OSHA Consultation Service.

If you elect not to use labor-management safety and health committees, be prepared to formalize and document your required system for communicating with employees.

Here are some helpful tips on complying with this difficult section:

1. Your communication system must be in a form "readily understandable by all affected employees." This means you should be prepared to communicate with employees in a language they can understand, and if an employee cannot read in any language, you must communicate with him/her orally in a language "readily understandable." Your communication system must be "designed to encourage employees to inform the employer of hazards at the workplace without fear of reprisal"; it must be a two-way system of communication.

2. Schedule general employee meetings at which safety is freely and openly discussed by those present. Such meetings should be regular, scheduled, and announced to all employees so that maximum employee attendance can be achieved. Remember to do this for all shifts. Many employers find it cost effective to hold such meetings at shift change time, with a brief overlap of schedules to accomplish the meetings. If properly planned, effective safety meetings can be held in a 15 to 20 minute time frame. Concentrate on:

   • Occupational accident and injury history at your own worksite, with possible comparisons to other locations in your company.

   • Feedback from the employee group.

   • Guest speakers from your worker's compensation insurance carrier or other agencies concerned with safety.

   • Brief audio-visual materials that relate to your industry.

   • Control of the meetings.

   • Stress that the purpose of the meeting is safety. Members of management should attend this meeting.

3. Training programs are excellent vehicles for communicating with employees.

4. Posters and bulletins can be very effective ways of communicating with employees. Useful materials can be obtained from Cal/OSHA, your workers' compensation insurance carrier, the National Safety Council or other commercial and public service agencies.

5. Newsletters or similar publications devoted to safety are also very effective communication devices. If you cannot devote resources to an entire publication, make safety a featured item in every issue of your company newsletter.

6. A safety suggestion box can be used by employees, anonymously if desired, to communicate their concerns to management.

7. Publish a brief company safety policy or statement informing all employees that safety is a priority issue with management, and urge employees to actively participate in the program for the common good of all concerned.

8. Communicate your concerns about safety to all levels of management.

9. Document all communication efforts, as you will be required to demonstrate that a system of effective communication is in place.

Hazard Assessment and Control

Periodic inspections and procedures for correction and control provide a method of identifying existing or potential hazards in the workplace and eliminating or controlling them. Hazard control is the heart of an effective Injury and Illness Prevention Program.

If hazards occur or recur, this reflects a breakdown in the hazard control system. The hazard control system is also the basis for developing safe work procedures and injury/illness prevention training.

The required hazard assessment survey of your establishment, when first developing your Injury and Illness Prevention Program, must be made by a qualified person. This survey can provide the basis and guide for establishing your hazard assessment and control system. The survey produces knowledge of hazards that exist in the workplace and conditions, equipment, and procedures that could be potentially hazardous.

An effective hazard control system will identify: hazards that exist or develop in your workplace, how to correct those hazards, and steps you can take to prevent their recurrence. If you have an effective system for monitoring workplace conditions:

1. You will be able to prevent many hazards from occurring through scheduled and documented self-inspections. Make sure established safe work practices are being followed and those unsafe conditions or procedures are identified and corrected properly. Scheduled inspections are in addition to the everyday safety and health checks that are part of the routine duties of managers and supervisors. The frequency of these inspections depends on the operations involved, the magnitude of the hazards, the proficiency of employees, changes in equipment or work processes, and the history of workplace injuries and illnesses. Inspections should be conducted by personnel who, through experience or training, are able to identify actual and potential hazards and understand safe work practices. Written inspection reports must be reviewed by management and/or the safety committee. The review should assist in prioritizing actions and verify completion of previous corrective actions. Overall inspection program results should be reviewed for trends.

2. Know which Cal/OSHA safety orders contained in Title 8 of the California Code of Regulations apply to your workplace and use them to identify potential hazards. A Cal/OSHA Consultation Service consultant or outside consultant can assist you in identifying safety orders applicable to your work.

3. Your employees should be encouraged to tell you or their supervisors of possibly hazardous situations, knowing their reports will be given prompt and serious attention without fear of reprisal. When you let them know that the situation was corrected (or why it was not hazardous), you create a system by which your employees continue to report hazards promptly and effectively.

4. Workplace equipment and personal, protective equipment should be maintained in safe and good working condition. In addition to what is required by Cal/ OSHA standards, your own program monitors the operation of workplace equipment and can also verify that routine preventive maintenance is conducted and personal protective equipment is reliable. This makes good safety sense, and proper maintenance can prevent costly breakdowns and undue exposures.

5. Hazards should be corrected as soon as they are identified. For any that can't be immediately corrected, set a target data for correction based on such considerations as the probability and severity of an injury or illness resulting from the hazard; the availability of needed equipment, materials, and/or personnel; time for delivery, installation, modification, or construction; and training periods. Provide interim protection to employees who need it while correction of hazards is proceeding. A written tracking system such as a log helps you monitor the progress of hazard correction.

6. You should review and prioritize your program based on the severity of the hazard.

Accident Investigation

A primary tool you should be using in an effort to identify and recognize the areas responsible for accidents is a thorough and properly completed accident investigation. It should be in writing and adequately identify the cause(s) of the accident or near-miss occurrence.

Accident investigations should be conducted by trained individuals, and with the primary focus of understanding why the accident or near miss occurred and what actions can be taken to preclude recurrence. In large organizations this responsibility may be assigned to the safety director. In smaller organizations the responsibility may lie directly with the supervisor responsible for the affected area or employee. Questions to ask in an accident investigation include:

1. **What happened?**

   The investigation should describe what took place that prompted the investigation: an injury to an employee, an incident that caused a production delay, damaged material, or any other conditions recognized as having a potential for losses or delays.

2. **Why did the incident happen?**

   The investigation must obtain all the facts surrounding the occurrence: what caused the situation to occur; who was involved; was/were the employee(s) qualified to perform the functions involved in the accident or near miss; were they properly trained; were proper operating procedures established for the task involved; were procedures followed, and if not, why not; where else this or a similar situation might exist, and how it can be corrected.

3. **What should be done?**

   The person conducting the investigation must determine which aspects of the operation or processes require additional attention. It is important to note that the purpose here is not to establish blame, but to determine what type of constructive action can eliminate the cause(s) of the accident or near miss.

4. **What action has been taken?**

   Action already taken to reduce or eliminate the exposures being investigated should be noted, along with those remaining to be addressed. Any interim or temporary precautions should also be noted. Any pending corrective action and reason for delaying its implementation should be identified.

Corrective action should be identified in terms of not only how it will prevent a recurrence of the accident or near miss but also how it will improve the overall operation. This will assist the investigator in selling his/her solutions to management. The solution should be a means of achieving not only accident control but also total operation control.

If you have a safety and health committee, its members should review investigations of all accidents and near-miss incidents to assist in recommending appropriate corrective actions to prevent a similar recurrence.

Thorough investigation of all accidents and near misses will help you identify causes and needed corrections and can help you determine why accidents occur, where they happen, and any accident trends. Such information is critical to preventing and controlling hazards and potential accidents.

Safety Planning, Rules, and Work Procedures

Planning for safety and health is an important part of every business decision, including purchasing, engineering, changes in work processes, and planning for emergencies. Your safety and health planning are effective when your workplace has:

1. Rules written to apply to everyone and addressing areas such as personal protective equipment, appropriate clothing, expected behavior, and emergency procedures. You and your employees should periodically review and update all rules and procedures to make sure they reflect present conditions. Rules and procedures should be written for new exposures when they are introduced into the workplace.

2. Safe and healthful work practices developed for each specific job.

3. Discipline or reward procedures to help assure that safety rules and work procedures are put into practice and enforced. Reward or positive reinforcement procedures such as bonus, incentive or employee recognition programs should provide positive motivation for compliance with safety rules and procedures.

4. A written plan for emergency situations. Your plan must include a list of emergencies that could arise and a set of procedures in response to each situation. Some emergency procedures, such as those covering medical emergencies or fire evacuation, are mandated by Cal/OSHA regulations.

5. If you have operations involving hazardous substances, procedures or processes, you must designate emergency response teams to be specifically trained and equipped to handle possible imminent hazards.

Safety and Health Training

Training is one of the most important elements of any Injury and Illness Prevention Program. It allows employees to learn their job properly, brings new ideas into the workplace, reinforces existing ideas and practices, and puts your program into action.

Your employees benefit from safety and health training through fewer work-related injuries and illnesses and reduced stress and worry caused by exposure to hazards.

You benefit from reduced workplace injuries and illnesses, increased productivity, lower costs, higher profits, and a more cohesive and dependable workforce.

An effective Injury and Illness Prevention Program includes training for both supervisors and employees. Training for both is required by Cal/OSHA safety orders.

You may need outside professionals to help you develop and conduct your required training program. Help is available from the Cal/OSHA Consultation Service, your workers' compensation insurance carrier, private consultants and vendor representatives.

Outside trainers should be considered temporary. Eventually you will need your own in-house training capabilities so you can provide training that is timely and specific to the needs of your workplace and your employees.

To be effective and also meet Cal/OSHA requirements, your training program needs to:

1. Let your supervisors know:
   - They are key figures responsible for establishment and success of your Injury and Illness Prevention Program.
   - The importance of establishing and maintaining safe and healthful working conditions.
   - They are responsible for being familiar with safety and health hazards to which their employees are exposed, how to recognize them, the potential effects these hazards have on the employees, and rules, procedures, and work practices for controlling exposure to those hazards.
   - How to convey this information to employees by setting good examples, instructing them, making sure they fully understand and follow safe procedures.
   - How to investigate accidents and take corrective and preventive action.
2. Let your employees know:
   - The success of the company's Injury and Illness Prevention Program depends on their actions as well as yours.
   - The safe work procedures required for their jobs and how these procedures protect them against exposure.
   - When personal protective equipment is required or needed, how to use it, and maintain it in good condition.
   - What to do if emergencies occur in the workplace.

An effective Injury and Illness Prevention Program requires proper job performance by everyone in the workplace. As the employer, you must ensure that all employees are knowledgeable about the materials and equipment they are working with, what known hazards are present, and how they are controlled.

Each employee needs to understand that:

- No employee is expected to undertake a job until he/she has received instructions on how to do it properly and safely and is authorized to perform the job.
- No employees should undertake a job that appears to be unsafe.
- No employee should use chemicals without fully understanding their toxic properties and without the knowledge required to work with them safely.
- Mechanical safeguards must always be in place and kept in place.
- Employees are to report to a superior or designated individual all unsafe conditions encountered during work.
- Any work-related injury or illness suffered, however slight, must be reported to management at once.
- Personal protective equipment must be used when and where required and properly maintained.

Your supervisors must recognize that they are the primary safety trainers in your organization. Encourage and help them by providing supervisory training. Many community colleges offer management training courses at little or no cost.

You as the employer are required under Cal/ OSHA standards to establish and carry out a formal training program. A professional training person, an outside consultant, or your supervisors may provide injury and illness prevention training to your employees.

This program must, at a minimum, provide training and instruction:

- To all employees when your program is first established.
- To all new employees.
- To all employees given new job assignments for which training has not been previously received.
- Whenever new substances, processes, procedures, or equipment are introduced to the workplace and present a new hazard.
- Whenever you or your supervisors are made aware of a new or previously unrecognized hazard.
- For all supervisors to assure they are familiar with the safety and health hazards to which employees under their immediate direction and control may be exposed.

The complete text of this Cal/OSHA guide can be found at http://www.dir.ca.gov/DOSH/.

Cal/OSHA has also developed sample Injury and Illness Prevention Program. For the construction industry, refer to the CAL/OSHA Model Injury and Illness Prevention Program for High-Hazard Industries in Resource D, located at the back of this book and at dewalt.com/guides.

CHAPTER 5

JOB HAZARD ANALYSIS

The job hazard analysis process has been around for many years. It may be called Job Hazard Analysis (JHA), Job Safety Analysis (JSA), Activity Hazard Analysis (AHA), or even Task Hazard Analysis (THA). Regardless of the title, the goal is the same: preventing workplace hazards and instructing workers on safe work practices. For the purpose of this guide, we will use the phrase job hazard analysis.

The JHA is an extension of a company safety and health program. Construction companies and contractors can help prevent many jobsite injuries and illness by performing a JHA. The purpose of a JHA is to break down each task performed on the job and analyze them. Once those tasks are analyzed, safe work practices are developed and then communicated to all employees.

Hazard awareness and hazard prevention can help you achieve your goal of zero injuries. A good starting point for hazard awareness is a job hazard analysis. The following information provided by OSHA relates to the analysis process.

OSHA JOB HAZARD ANALYSIS GUIDE

The following are excerpts from OSHA Publication 3071. This document describes the process of the JHA. The full version of this publication can be found at http://www.osha.gov.[1]

What Is a Hazard?

A hazard is the potential for harm. In practical terms, a hazard often is associated with a condition or activity that, if left uncontrolled, can result in an injury or illness. [See Resource E, located at the back of this book and at dewalt.com/guides, for a list of common hazards and descriptions.] Identifying hazards and eliminating or controlling them as early as possible will help prevent injuries and illnesses.

What Is a Job Hazard Analysis?

A job hazard analysis is a technique that focuses on job tasks as a way to identify hazards before they occur. It focuses on the relationship between the worker, the task, the tools, and the work environment. Ideally, after you identify uncontrolled hazards, you will take steps to eliminate or reduce them to an acceptable risk level.

Why Is Job Hazard Analysis Important?

Many workers are injured and killed at the workplace every day in the United States. Safety and health can add value to your business, your job, and your life. You can help prevent workplace injuries and illnesses by looking at your workplace operations, establishing proper job procedures, and ensuring that all employees are trained properly.

---

[1]Occupational Safety & Health Administration, U.S. Department of Labor. *Job Hazard Analysis*, OSHA Publication 3071, revised 2002. http://www.osha.gov.

One of the best ways to determine and establish proper work procedures is to conduct a job hazard analysis. A job hazard analysis is one component of the larger commitment of a safety and health management system.

What Is the Value of a Job Hazard Analysis?

Supervisors can use the findings of a job hazard analysis to eliminate and prevent hazards in their workplaces. This is likely to result in fewer worker injuries and illnesses; safer, more effective work methods; reduced workers' compensation costs; and increased worker productivity. The analysis also can be a valuable tool for training new employees in the steps required to perform their jobs safely.

For a job hazard analysis to be effective, management must demonstrate its commitment to safety and health and follow through to correct any uncontrolled hazards identified. Otherwise, management will lose credibility and employees may hesitate to go to management when dangerous conditions threaten them.

What Jobs Are Appropriate for a Job Hazard Analysis?

A job hazard analysis can be conducted on many jobs in your workplace. Priority should go to the following types of jobs:

- Jobs with the highest injury or illness rates;
- Jobs with the potential to cause severe or disabling injuries or illness, even if there is no history of previous accidents;
- Jobs in which one simple human error could lead to a severe accident or injury;
- Jobs that are new to your operation or have undergone changes in processes and procedures; and
- Jobs complex enough to require written instructions.

Where Do I Begin?

1. **Involve your employees.** It is very important to involve your employees in the hazard analysis process. They have a unique understanding of the job, and this knowledge is invaluable for finding hazards. Involving employees will help minimize oversights, ensure a quality analysis, and get workers to "buy in" to the solutions because they will share ownership in their safety and health program.

2. **Review your accident history.** Review with your employees your worksite's history of accidents and occupational illnesses that needed treatment, losses that required repair or replacement, and any "near misses"—events in which an accident or loss did not occur, but could have. These events are indicators that the existing hazard controls (if any) may not be adequate and deserve more scrutiny.

3. **Conduct a preliminary job review.** Discuss with your employees the hazards they know exist in their current work and surroundings. Brainstorm with them for ideas to eliminate or control those hazards.

    If any hazards exist that pose an immediate danger to an employee's life or health, take immediate action to protect the worker. Any problems that can be corrected easily should be corrected as soon as possible. Do not wait to complete your job hazard analysis. This will demonstrate your commitment to safety and health and enable you to focus on the hazards and jobs that need more study because of their complexity. For those hazards determined to present unacceptable risks, evaluate types of hazard controls.

4. **List, rank, and set priorities for hazardous jobs.** List jobs with hazards that present unacceptable risks, based on those most likely to occur and with the most severe consequences. These jobs should be your first priority for analysis.

5. **Outline the steps or tasks.** Nearly every job can be broken down into job tasks or steps. When beginning a job hazard analysis, watch the employee perform the job and list each step as the worker takes it. Be sure to record enough information to describe each job action without getting overly detailed. Avoid making the breakdown of steps so detailed that it becomes unnecessarily long or so broad that it does not include basic steps. You may find it valuable to get input from other workers who have performed the same job. Later, review the job steps with the employee to make sure you have not omitted something. Point out that you are evaluating the job itself, not the employee's job performance. Include the employee in all phases of the analysis—from reviewing the job steps and procedures to discussing uncontrolled hazards and recommended solutions.

Sometimes, in conducting a job hazard analysis, it may be helpful to photograph or videotape the worker performing the job. These visual records can be handy references when doing a more detailed analysis of the work.

How Do I Identify Workplace Hazards?

A job hazard analysis is an exercise in detective work. Your goal is to discover the following:

- What can go wrong?
- What are the consequences?
- How could it happen?
- What are other contributing factors?
- How likely is it that the hazard will occur?

To make your job hazard analysis useful, document the answers to these questions in a consistent manner. Describing a hazard in this way helps to ensure that your efforts to eliminate the hazard and implement hazard controls help target the most important contributors to the hazard.

Good hazard scenarios describe:

- Where it is happening (environment),
- Who or what it is happening to (exposure),
- What precipitates the hazard (trigger),
- The outcome that would occur should it happen (consequence), and
- Any other contributing factors.

[A sample JHA form found in Resource F, located at the back of this book, helps you organize your information to provide these details. This form can also be found at dewalt.com/guides.]

Rarely is a hazard a simple case of one singular cause resulting in one singular effect. More frequently, many contributing factors tend to line up in a certain way to create the hazard. Here is an example of a hazard scenario: In the metal shop (environment), while clearing a snag (trigger), a worker's hand (exposure) comes into contact with a rotating pulley. It pulls his hand into the machine and severs his fingers (consequences) quickly.

To perform a job hazard analysis, you would ask:

What can go wrong?	The worker's hand could come into contact with a rotating object that "catches" it and pulls it into the machine.
What are the consequences?	The worker could receive a severe injury and lose fingers and hands.
How could it happen?	The accident could happen as a result of the worker trying to clear a snag during operations or as part of a maintenance activity while the pulley is operating. Obviously, this hazard scenario could not occur if the pulley is not rotating.
What are other contributing factors?	This hazard occurs very quickly. It does not give the worker much opportunity to recover or prevent it once his hand comes into contact with the pulley. This is an important factor, because it helps you determine the severity and likelihood of an accident when selecting appropriate hazard controls. Unfortunately, experience has shown that training is not very effective in hazard control when triggering events happen quickly because humans can react only so quickly.
How likely is it that the hazard will occur?	This determination requires some judgment. If there have been "near misses" or actual cases, then the likelihood of a recurrence would be considered high. If the pulley is exposed and easily accessible, that also is a consideration. In the example, the likelihood that the hazard will occur is high because there is no guard preventing contact, and the operation is performed while the machine is running. By following the steps in this example, you can organize your hazard analysis activities.

The following example shows how a job hazard analysis can be used to identify the existing or potential hazards for each basic step involved in grinding iron castings.

Grinding Iron Castings: Job Steps

Step 1. Reach into the metal box to the right of the machine, grasp the casting, and carry it to the wheel.

Step 2. Push the casting against the wheel to grind off any burr.

Step 3. Place the finished casting in the box to the left of the machine.

Job Location: Metal Shop	
Analyst: Joe Safety	
Date:	**Time:**
Task Description	The worker reaches into the metal box to the right of the machine, grasps a 15-pound casting, and carries it to the grinding wheel. The worker grinds 20 to 30 castings per hour.
Hazard Description	Picking up a casting, the employee could drop it onto his foot. The casting's weight and height could seriously injure the worker's foot or toes.
Hazard Controls	1. Remove the castings from the box and place them on a table next to the grinder. 2. Wear steel-toed shoes with arch protection. 3. Change to protective gloves that allow a better grip. 4. Use a device to pick up the castings.
Task Description	The worker reaches into the metal box to the right of the machine, grasps a 15-pound casting, and carries it to the grinding wheel. The worker grinds 20 to 30 castings per hour.
Hazard Description	Castings have sharp burrs and edges that can cause severe lacerations.
Hazard Controls	1. Use a device such as a clamp to pick up the castings. 2. Wear cut-resistant gloves that allow a good grip and fit tightly to minimize the chance that they will get caught in the grinding wheel.
Task Description	The worker reaches into the metal box to the right of the machine, grasps a 15-pound casting, and carries it to the grinding wheel. The worker grinds 20 to 30 castings per hour.
Hazard Description	Reaching, twisting, and lifting 15-pound castings from the floor could result in a muscle strain to the lower back.
Hazard Controls	1. Move the castings from the ground and place them closer to the work zone to minimize lifting. Ideally, place them at waist height or on an adjustable platform or pallet. 2. Train workers not to twist while lifting and reconfigure work stations to minimize twisting during lifts.

How Do I Correct or Prevent Hazards?

After reviewing your list of hazards with the employee, consider what control methods will eliminate or reduce them. [For more information on JHA hazard control measures, see Resource G located at the back of this book and at dewalt.com/guides.] The most effective controls are engineering controls that physically change a machine or work environment to prevent employee exposure to the hazard. The more reliable a hazard control or the less likely a hazard control can be circumvented, the better. If this is not feasible, administrative controls may be appropriate. This may involve changing how employees do their jobs.

Discuss your recommendations with all employees who perform the job and consider their responses carefully. If you plan to introduce new or modified job procedures, be sure employees understand what they are required to do and the reasons for the changes.

What Else Do I Need to Know Before Starting a Job Hazard Analysis?

The job procedures discussed in this booklet are for illustration only and do not necessarily include all the steps, hazards, and protections that apply to your industry. When conducting your own job safety analysis, be sure to consult the Occupational Safety and Health Administration standards for your industry. Compliance with these standards is mandatory, and by incorporating their requirements in your job hazard analysis, you can be sure that your health and safety program meets federal standards.

Why Should I Review My Job Hazard Analysis?

Periodically reviewing your job hazard analysis ensures that it remains current and continues to help reduce workplace accidents and injuries. Even if the job has not changed, it is possible that during the review process you will identify hazards that were not identified in the initial analysis.

It is particularly important to review your job hazard analysis if an illness or injury occurs on a specific job. Based on the circumstances, you may determine that you need to change the job procedure to prevent similar incidents in the future. If an employee's failure to follow proper job procedures results in a "close call," discuss the situation with all employees who perform the job and remind them of proper procedures. Any time you revise a job hazard analysis, it is important to train all employees affected by the changes in the new job methods, procedures, or protective measures adopted.

When Is It Appropriate to Hire a Professional to Conduct a Job Hazard Analysis?

If your employees are involved in many different or complex processes, you need professional help conducting your job hazard analyses. Sources of help include your insurance company, the local fire department, and private consultants with safety and health expertise. In addition, OSHA offers assistance through its regional and area offices and consultation services.

Even when you receive outside help, it is important that you and your employees remain involved in the process of identifying and correcting hazards because you are on the worksite every day and most likely to encounter these hazards. New circumstances and a recombination of existing circumstances may cause old hazards to reappear and new hazards to appear. In addition, you and your employees must be ready and able to implement whatever hazard elimination or control measures a professional consultant recommends.

CHAPTER 6

SAFETY AND HEALTH TRAINING

Safety training is a vital part of any organization's safety and health program. Construction contractors need to ensure that they train all of their employees how to perform their jobs safely. Unfortunately, many contractors do not provide adequate safety training for their employees. You would not hire an inexperienced person to pull electrical wires for your company without training them on how to pull wires. Why would you hire an employee and not train them how to perform their job safely? OSHA regulation 1926.21(b)(2) requires employers to train their employees.[1]

The employer shall instruct each employee in the recognition and avoidance of unsafe conditions and the regulations applicable to his work environment to control or eliminate any hazards or other exposure to illness or injury.

All employees should therefore be instructed in safe work practices concerning all jobsite hazards they would be expected to encounter during their jobs. Employers have many options for training their workers, including, but not limited to, new employee orientation, on-the-job training, task-specific training (training based on a JHA), safety meetings, and OSHA 10- or 30-hour training.

OSHA TRAINING GUIDELINES FOR THE CONSTRUCTION INDUSTRY

OSHA has developed training guidelines to assist contractors in the construction industry, which can be found in OSHA Publication 2254. The full publication can be found at http://www.osha.gov. The following sections are excerpts from OSHA's guide[2]:

Training Guidelines

OSHA's training guidelines follow a model that consists of:

Determining If Training Is Needed	Identifying Training Needs
Identifying Goals and Objectives	Developing Learning Activities
Conducting the Training	Evaluating Program Effectiveness
Improving the Program	Retraining and Continuous Education

[1]Occupational Safety & Healthy Administration, U.S. Department of Labor. *Safety and Health Regulations for Construction*, Standard Number 1926.21(b)(2). http://www.osha.gov.
[2]Occupational Safety & Health Administration, U.S. Department of Labor. *Training Requirements in OSHA Standards and Training Guidelines*, OSHA 2254, 1998 (Revised). http://www.osha.gov.

Determining If Training Is Needed

The first step in the training process is a basic one: to determine whether a problem can be solved by training. Whenever employees are not performing their jobs properly, it is often assumed that training will bring them up to standard. However, it is possible that other actions (such as hazard abatement or the implementation of engineering controls) would enable employees to perform their jobs properly.

Ideally, safety and health training should be provided before problems or accidents occur. This training would cover both general safety and health rules and work procedures, and would be repeated if an accident or near-miss incident occurred. Problems that can be addressed effectively by training include those that arise from lack of knowledge of a work process, unfamiliarity with equipment, or incorrect execution of a task. Training is less effective (but still can be used) for problems arising from an employee's lack of motivation or lack of attention to the job. Whatever its purpose, training is most effective when designed in relation to the goals of the employer's total safety and health program.

Identifying Training Needs

If the problem is one that can be solved, in whole or in part, by training, then the next step is to determine what training is needed. For this, it is necessary to identify what the employee is expected to do and in what ways, if any, the employee's performance is deficient. This information can be obtained by conducting a job analysis which pinpoints what an employee needs to know in order to perform a job.

When designing a new training program, or preparing to instruct an employee in an unfamiliar procedure or system, a JHA can be developed by examining engineering data on new equipment or the material safety data sheets on unfamiliar substances. The content of the specific Federal or State OSHA standards applicable to a business can also provide direction in developing training content. Another option is to conduct a Job Hazard Analysis. This is a procedure for studying and recording each step of a job, identifying existing or potential hazards, and determining the best way to perform the job in order to reduce or eliminate the risks. Information obtained from a JHA can be used as the content for the training activity. [JHAs are covered in depth in Chapter 5.]

If an employer's learning needs can be met by revising an existing training program rather than developing a new one, or if the employer already has some knowledge of the process or system to be used, appropriate training content can be developed through such means as:

- Using company accident and injury records to identify how accidents occur and what can be done to prevent them from recurring.

- Requesting employees to provide, in writing and in their own words, descriptions of their jobs. These should include the tasks performed and the tools, materials, and equipment used.

- Observing employees at the worksite as they perform tasks, asking about the work, and recording their answers.

- Examining similar training programs offered by other companies in the same industry, or obtaining suggestions from such organizations as the National Safety Council, the Bureau of Labor Statistics, OSHA-approved State programs, OSHA full-service Area Offices, OSHA-funded State consultation programs, or the OSHA Office of Training and Education.

The employees themselves can provide valuable information on the training they need. Safety and health hazards can be identified through the employees' responses to such questions as whether anything about their jobs frightens them, if they have had any near-miss incidents, if they feel they are taking risks, or if they believe that their jobs involve hazardous operations or substances.

Identifying Goals and Objectives

Once the kind of training that is needed has been determined, it is equally important to determine what kind of training is not needed. Employees should be made aware of all the steps involved in a task or procedure, but training should focus on those steps on which improved performance is needed. This avoids unnecessary training and tailors the training to meet the needs of the employees.

Once the employees' training needs have been identified, employers can then prepare objectives for the training. Instructional objectives, if clearly stated, will tell employers what they want their employees to do, to do better, or to stop doing.

Learning objectives do not necessarily have to be written, but in order for the training to be as successful as possible, clear and measurable objectives should be thought out before the training begins. For an objective to be effective it should identify as precisely as possible what the individuals will do to demonstrate that they have learned or that the objective has been reached. They should also describe the important conditions under which the individual will demonstrate competence and define what constitutes acceptable performance.

Using specific, action-oriented language, the instructional objectives should describe the preferred practice or skill and its observable behavior. For example, rather than using the statement: "The employee will understand how to use a respirator" as an instructional objective, it would be better to say: "The employee will be able to describe how a respirator works and when it should be used." Objectives are most effective when worded in sufficient detail that other qualified persons can recognize when the desired behavior is exhibited.

Developing Learning Activities

Once employers have stated precisely what the objectives for the training program are, then learning activities can be identified and described. Learning activities enable employees to demonstrate that they have acquired the desired skills and knowledge. To ensure that employees transfer the skills or knowledge from the learning activity to the job, the learning situation should simulate the actual job as closely as possible. Thus, employers may want to arrange the objectives and activities in a sequence which corresponds to the order in which the tasks are to be performed on the job, if a specific process is to be learned. For instance, if an employee must learn the beginning processes of using a table saw, the sequence might be:

1. to check that the power source is connected;
2. to ensure that the safety devices are in place and are operative;
3. to know when and how to throw the switch to turn it on; and so on.

A few factors will help to determine the type of learning activity to be incorporated into the training. One aspect is the training resources available to the employer. Can a group training program that uses an outside trainer and film be organized, or should the employer personally train the employees on a one-to-one basis? Another factor is the kind of skills or knowledge to be learned. Is the learning oriented toward physical skills (such as the use of special tools) or toward mental processes and attitudes? Such factors will influence the type of learning activity designed by employers. The training activity can be group-oriented, with lectures, role play, and demonstrations; or designed for the individual as with self-paced instruction.

The determination of methods and materials for the learning activity can be as varied as the employer's imagination and available resources will allow. The employer may want to use charts, diagrams, manuals, slides, films, viewgraphs (overhead transparencies), videotapes, audiotapes, or simply blackboard and chalk, or any combination of these and other instructional aides. Whatever the method of instruction, the learning activities should be developed in such a way that the employees can clearly demonstrate that they have acquired the desired skills or knowledge.

Conducting the Training

With the completion of the steps outlined above, the employer is ready to begin conducting the training. To the extent possible, the training should be presented so that its organization and meaning are clear to the employees. To do so, employers or supervisors should:

1. provide overviews of the material to be learned;
2. relate, wherever possible, the new information or skills to the employee's goals, interests, or experience; and
3. reinforce what the employees learned by summarizing the program's objectives and the key points of information covered.

These steps will assist employers in presenting the training in a clear, unambiguous manner.

In addition to organizing the content, employers must also develop the structure and format of the training. The content developed for the program, the nature of the workplace or other training site, and the resources available for training will help employers determine for themselves the frequency of training activities, the length of the sessions, the instructional techniques, and the individual(s) best qualified to present the information.

In order to be motivated to pay attention and learn the material that the employer or supervisor is presenting, employees must be convinced of the importance and relevance of the material. Among the ways of developing motivation are:

1. explaining the goals and objectives of instruction;
2. relating the training to the interests, skills, and experiences of the employees;

3. outlining the main points to be presented during the training session(s); and

4. pointing out the benefits of training (e.g., the employee will be better informed, more skilled, and thus more valuable both on the job and on the labor market; or the employee will, if he or she applies the skills and knowledge learned, be able to work at reduced risk).

An effective training program allows employees to participate in the training process and to practice their skills or knowledge. This will help to ensure that they are learning the required knowledge or skills and permit correction if necessary. Employees can become involved in the training process by participating in discussions, asking questions, contributing their knowledge and expertise, learning through hands-on experiences, and through role-playing exercises.

Evaluating Program Effectiveness

To make sure that the training program is accomplishing its goals, an evaluation of the training can be valuable. Training should have, as one of its critical components, a method of measuring the effectiveness of the training. A plan for evaluating the training session(s), either written or thought out by the employer, should be developed when the course objectives and content are developed. It should not be delayed until the training has been completed. Evaluation will help employers or supervisors determine the amount of learning achieved and whether an employee's performance has improved on the job. Among the methods of evaluating training are:

1. **Student opinion.** Questionnaires or informal discussions with employees can help employers determine the relevance and appropriateness of the training program;

2. **Supervisors' observations.** Supervisors are in good positions to observe an employee's performance both before and after the training and note improvements or changes; and

3. **Workplace improvements.** The ultimate success of a training program may be changes throughout the workplace that result in reduced injury or accident rates.

However it is conducted, an evaluation of training can give employers the information necessary to decide whether or not the employees achieved the desired results and whether the training session should be offered again at some future date.

Improving the Program

If, after evaluation, it is clear that the training did not give the employees the level of knowledge and skill that was expected, then it may be necessary to revise the training program or provide periodic retraining. At this point, asking questions of employees and of those who conducted the training may be of some help. Among the questions that could be asked are:

1. Were parts of the content already known and, therefore, unnecessary?

2. What material was confusing or distracting?

3. Was anything missing from the program?

4. What did the employees learn, and what did they fail to learn?

It may be necessary to repeat steps in the training process, that is, to return to the first steps and retrace one's way through the training process. As the program is evaluated, the employer should ask:

1. If a JHA was conducted, was it accurate?

2. Was any critical feature of the job overlooked?

3. Were the important gaps in knowledge and skill included?

4. Was material already known by the employees intentionally omitted?

5. Were the instructional objectives presented clearly and concretely?

6. Did the objectives state the level of acceptable performance that was expected of employees?

7. Did the learning activity simulate the actual job?

8. Was the learning activity appropriate for the kinds of knowledge and skills required on the job?

9. When the training was presented, was the organization of the material and its meaning made clear?

10. Were the employees motivated to learn?

11. Were the employees allowed to participate actively in the training process?

12. Was the employer's evaluation of the program thorough?

A critical examination of the steps in the training process will help employers to determine where course revision is necessary.

Matching Training to Employees

While all employees are entitled to know as much as possible about the safety and health hazards to which they are exposed, and employers should attempt to provide all relevant information and instruction to all employees, the resources for such an effort frequently are not, or are not believed to be, available. Thus, employers are often faced with the problem of deciding who is in the greatest need of information and instruction.

One way to differentiate between employees who have priority needs for training and those who do not is to identify employee populations which are at higher levels of risk. The nature of the work will provide an indication that such groups should receive priority for information on occupational safety and health risks.

Identifying Employees at Risk

One method of identifying employee populations at high levels of occupational risk (and thus in greater need of safety and health training) is to pinpoint hazardous occupations. Even within industries which are hazardous in general, there are some employees who operate at greater risk than others. In other cases the hazardousness of an occupation is influenced by the conditions under which it is performed, such as noise, heat or cold, or safety or health hazards in the surrounding area. In these situations, employees should be trained not only on how to perform their job safely but also on how to operate within a hazardous environment.

A second method of identifying employee populations at high levels of risk is to examine the incidence of accidents and injuries, both within the company and within the industry. If employees in certain occupational categories are experiencing higher accident and injury rates than other employees, training may be one way to reduce that rate. In addition, thorough accident investigation can identify not only specific employees who could benefit from training but also identify company-wide training needs.

Research has identified the following variables as being related to a disproportionate share of injuries and illnesses at the worksite on the part of employees:

- The age of the employee (younger employees have higher incidence rates).
- The length of time on the job (new employees have higher incidence rates).
- The size of the firm (in general terms, medium-size firms have higher incidence rates than smaller or larger firms).
- The type of work performed (incidence and severity rates vary significantly by SIC Code).
- The use of hazardous substances (by SIC Code).

These variables should be considered when identifying employee groups for training in occupational safety and health.

In summary, information is readily available to help employers identify which employees should receive safety and health information, education and training, and who should receive it before others. Employers can request assistance in obtaining information by contacting such organizations as OSHA Area Offices, the Bureau of Labor Statistics, OSHA-approved State programs, State onsite consultation programs, the OSHA Office of Training and Education, or local safety councils.

Training Employees at Risk

Determining the content of training for employee populations at higher levels of risk is similar to determining what any employee needs to know, but more emphasis is placed on the requirements of the job and the possibility of injury.

One useful tool for determining training content from job requirements is the Job Hazard Analysis described earlier. This procedure examines each step of a job, identifies existing or potential hazards, and determines the best way to perform the job in order to reduce or eliminate the hazards. Its key elements are:

1. job description;
2. job location;
3. key steps (preferably in the order in which they are performed);
4. tools, machines, and materials used;
5. actual and potential safety and health hazards associated with these key job steps; and
6. safe and healthful practices, apparel, and equipment required for each job step.

Material Safety Data Sheets (MSDS) can also provide information for training employees in the safe use of materials. These data sheets, developed by chemical manufacturers and importers, are supplied with manufacturing or construction materials and describe the ingredients of a product, its hazards, protective equipment to be used, safe handling procedures, and emergency first-aid responses. The information contained in these sheets can help employers identify employees in need of training (i.e., workers handling substances described in the sheets) and train employees in safe use of the substances. Material Safety Data Sheets are generally available from suppliers, manufacturers of the substance, large employers who use the substance on a regular basis, or they can be developed by employers or trade associations. MSDS are particularly useful for those employers who are developing training on chemical use as required by OSHA's Hazard Communication Standard.

OSHA 10- AND 30-HOUR TRAINING REQUIREMENTS

OSHA developed an outreach training program that offers OSHA 10-hour and 30-hour training. The training offers a solid base of safety knowledge for the construction industry and is completely voluntary for federal OSHA.

There are a few state-run OSHA plans that have regulatory requirements for OSHA 10- and/or 30-hour training. Employers should check with the state where they work to determine if it has those requirements. A few states that do have the requirements are listed below:

Nevada	In 2010, a state law became effective requiring construction workers to have either an OSHA 10-hour construction safety course or OSHA 30-hour construction safety course. The 10-hour course is required for all nonsupervisory employees on the jobsite, and the OSHA 30-hour course is required for all supervisory employees on the jobsite. These training courses are valid for 5 years and must be retaken after that time.
New York	Requires OSHA 10-hour construction safety courses to be provided for all employees working on public works projects valued at $250,000 or greater.
Missouri	Requires OSHA 10-hour construction safety courses to be provided for all employees working on public works projects.
Massachusetts	Requires OSHA 10-hour construction safety courses for certain employees.
Rhode Island	Requires OSHA 10-hour construction safety courses for certain employees.
Connecticut	Requires OSHA 10-hour construction safety courses for certain employees.
New Hampshire	Requires OSHA 10-hour construction safety courses for certain employees.

ONLINE SAFETY TRAINING

A newer option to traditional classroom training is web-based training. Employers can provide their employees safety training specific to their industry to facilitate compliance with the OSHA regulations. Various companies offer online training courses.

SAFETY MEETINGS/TOOL BOX TALKS

Construction companies need to ensure that the safety and health training they provide their employees continues beyond the traditional classroom or online training. One method for continuing this training is holding safety meetings regularly. Safety meetings, also known as Tool Box Talks, should be designed to present a safety topic on a weekly or monthly basis to ensure that training is continual. In addition to presenting a safety topic, safety meetings can be a great way to increase employee involvement in your safety and health program. Safety meetings can actually be planned and presented by employees based on the topics that they feel are the most pressing.

How to Conduct a Safety Meeting

Find a Topic	There are many sources for safety meeting topics. A simple search of the Internet will reveal many free topics. Take time to find a topic that is relevant to your company and operation. A few sample safety meeting topics are presented in Resource H, located at the back of this book and at dewalt.com/guides.
Prepare	You can present safety topics by simply reading them, but meetings will be much more effective if you spend preparation time. Once you have picked a subject, add your own ideas. Use examples from the jobsite, and emphasize areas that are central in your circumstances. The safety meeting should be designed to last about 10 minutes; however, they can be longer when there are questions and discussion.
Pick a Location	The most common location is the construction trailer or office. This may be a great place for most meetings, but don't hesitate to change the locations for a more effective meeting. If you will be discussing ladders, you may want to set up a ladder and conduct the meeting at that location. Keep in mind that in order to maintain focus, avoid meeting in areas with distractions.
Pick a Date and Time	It is good practice to pick a consistent day and time each week or month. This will allow the attendees to schedule the meeting in their calendars in advance and will increase attendance. Many construction companies hold their meetings every Monday morning. This allows all workers to start the week off with a positive safety message. It is important to hold these meetings regularly. Safety is a state of mind and regular reminders can make workers more conscious of dangers.
Tips for Conducting the Meeting	• Always remind your employees that safe work practices are only effective if correctly implemented. The attitudes and work habits of everyone on the jobsite can have a huge impact on safety. • Let the attendees know in advance that the safety meeting will last about 15 minutes. • Introduce the safety topic in your own words and explain why it is relevant to your circumstances. Present the main ideas of the safety topic and ask questions to develop discussion. • After you present the safety topic, continue to encourage discussion among employees about the subject. At this time, reviewing recent jobsite accidents or incidents will further the discussion. It is then appropriate to ask for suggestions about how the accident or incident could have been prevented. If there were no accidents or incidents, congratulate the workers and then cover topics from a recent safety inspection. Be sure not to criticize anyone by name in front of the group.

(continued)

	• Encourage employees to recall all incidents or near misses. Try to get the group to learn from these experiences.
Document the Meeting	All attendees of the meeting should sign a safety meeting roster for proper documentation. Documentation will be presented in more detail in Chapter 12. A sample safety meeting roster can be found in Resource I, located at the back of this book and at dewalt.com/guides.

Having a well-trained team can go a long way toward creating a safe workplace. Remember, accidents are costly, but talk is cheap. Invest more time in safety and health training to ensure that employees are given the best opportunity to work safely.

CHAPTER 7

OSHA REGULATIONS AND SAFE WORK PRACTICES

This guide is intended to provide an overview of particular OSHA regulations and safe work practices. It is not an exhaustive guide to OSHA regulations; instead, it is intended to be a plain-language explanation of some regulations. The regulations are set forth in CFR Part 1926 and the OSH Act. OSHA's interpretations and enforcement policies may change over time. For additional guidance on OSHA compliance requirements, the reader should consult current administrative interpretations and decisions by the Occupational Safety and Health Review Commission and the courts.

Employers and employees in the states and territories that operate their own OSHA plans should check with their state agency. Their state may be enforcing standards and other procedures that, while "at least as effective as" federal standards, are not always identical to the federal requirements.

GENERAL DUTY CLAUSE

Chapter 1 presented the General Duty Clause, which is contained in the OSH Act. As a reminder, this clause allows OSHA to cite construction companies even if there is no specific regulation for the hazard being cited. If the hazard is recognized and likely to cause serious injury or death, OSHA will use the General Duty Clause[1]:

5. Duties
  (a) Each employer -
     (1) shall furnish to each of his employees employment and a place of employment which are free from recognized hazards that are causing or are likely to cause death or serious physical harm to his employees;

The General Duty Clause may seem difficult to comply with; however, if construction companies create a safety and health program that incorporates a job hazard analysis, they will find hazards and create safe work practices to deal with those hazards, thus reducing or eliminating exposure to the hazard.

GENERAL SAFETY AND HEALTH PROVISIONS

Contractor Requirements

No contractor or subcontractor for any part of contract work for construction, alteration, and/or repair, including painting and decorating, shall require any employee to work in surroundings or under working conditions which are unsanitary, hazardous, or dangerous to his/her health or safety.

---

[1] Occupational Safety & Health Administration, U.S. Department of Labor. *General Duty Clause*. Section 5(a)(1) OSH Act. http://www.osha.gov.

Accident Prevention Responsibilities

It shall be the responsibility of the employer to provide for frequent and regular inspections of the job site, materials, and equipment by competent persons designated by the employer.

A competent person is defined in §1926.32(f) as "one who is capable of identifying existing and predictable hazards in the surroundings or working conditions which are unsanitary, hazardous, or dangerous to employees, and who has authorization to take prompt corrective measures to eliminate them."

Unsafe Tools and Machinery

The use of any machinery, tool, material, or equipment which in not in compliance with OSHA standards is prohibited and they shall be identified as unsafe by tagging or locking the controls to render them inoperable or shall be physically removed from their place of operation.

Use of Equipment and Machinery

The employer shall permit only those employees qualified by training or experience to operate equipment and machinery.

Safety Training and Education

The employer shall instruct each employee in the recognition and avoidance of unsafe conditions and the regulations applicable to his work environment to control or eliminate any hazards or other exposure to illness or injury.

Employees required to handle or use poisons, caustics, and other harmful substances shall be instructed regarding the safe handling and use, and be made aware of the potential hazards, personal hygiene, and personal protective measures required.

All employees required to enter into confined or enclosed spaces shall be instructed as to the nature of the hazards involved, the necessary precautions to be taken, and in the use of protective and emergency equipment required. The employer shall comply with any specific regulations that apply to work in dangerous or potentially dangerous areas. OSHA defines "confined or enclosed space" as[2]:

> Any space having a limited means of egress, which is subject to the accumulation of toxic or flammable contaminants or has an oxygen deficient atmosphere. Confined or enclosed spaces include, but are not limited to, storage tanks, process vessels, bins, boilers, ventilation or exhaust ducts, sewers, underground utility vaults, tunnels, pipelines, and open top spaces more than 4 feet in depth such as pits, tubs, vaults, and vessels.

Confined Spaces

A confined space has limited openings for entry or exit, is large enough for entering and working, and is not designed for continuous worker occupancy. Confined spaces include underground vaults, tanks, storage bins, manholes, pits, silos, underground utility vaults and pipelines.

Permit-required confined spaces are confined spaces that:

- May contain a hazardous or potentially hazardous atmosphere
- May contain a material which can engulf an entrant
- May contain walls that converge inward or floors that slope downward and taper into a smaller area which could trap or asphyxiate an entrant
- May contain other serious physical hazards such as unguarded machines or exposed live wires

[2] Occupational Safety & Health Administration, U.S. Department of Labor. *Construction Safety and Health Regulation*. Subpart C, Section 1926.1. http://www.osha.gov.

- Must be identified by the employer who must inform exposed employees of the existence and location of such spaces and their hazards

What you should do:

- Do not enter permit-required confined spaces without being trained and without having a permit to enter
- Review, understand, and follow employer's procedures before entering permit-required confined spaces and know how and when to exit
- Before entry, identify any physical hazards
- Before and during entry, test and monitor for oxygen content, flammability, toxicity, or explosive hazards as necessary
- Use employer's fall protection, rescue, air monitoring, ventilation, lighting, and communication equipment according to entry procedures
- Maintain contact at all times with a trained attendant either visually, via phone, or by two-way radio. This monitoring system enables the attendant and entry supervisor to order you to evacuate and to alert appropriately trained rescue personnel to rescue entrants when needed

Housekeeping

During the course of construction, alteration, or repairs, form and scrap lumber with protruding nails, and all other debris, shall be kept cleared from work areas, passageways, and stairs, in and around buildings or other structures.

Combustible scrap and debris shall be removed at regular intervals during the course of construction. Safe means shall be provided to facilitate such removal.

Containers shall be provided for the collection and separation of waste, trash, oily and used rags, and other refuse. Containers used for garbage and other oily, flammable, or hazardous wastes, such as caustics, acids, harmful dusts, etc., shall be equipped with covers. Garbage and other waste shall be disposed of at frequent and regular intervals.

Means of Egress

In every building or structure, exits shall be so arranged and maintained as to provide free and unobstructed egress from all parts of the building or structure at all times when it is occupied. No lock or fastening to prevent free escape from the inside of any building shall be installed except in mental, penal, or corrective institutions where supervisory personnel are continually on duty and effective provisions are made to remove occupants in case of fire or other emergency.

Exits shall be marked by a readily visible sign. Access to exits shall be marked by readily visible signs in all cases where the exit or way to reach it is not immediately visible to the occupants.

Means of egress shall be continually maintained free of all obstructions or impediments to full instant use in the case of fire or other emergency.

OCCUPATIONAL HEALTH AND ENVIRONMENTAL CONTROLS

Medical Services and First Aid

The employer shall ensure the availability of medical personnel for advice and consultation on matters of occupational health. When a medical facility is not reasonably accessible for the treatment of injured employees, a person qualified to render first aid shall be available at the worksite. It is recommended that more than one person on site is qualified to render first aid.

Adequate first-aid supplies should be readily available. In areas where 911 is not available, the telephone numbers of the physicians, hospitals, or ambulances shall be conspicuously posted and the address and directions to the closest hospital should be made available to workers.

Drinking Water

An adequate supply of potable water shall be provided in all places of employment. Portable drinking water containers shall be capable of being tightly closed and equipped with a tap. Using a common drinking cup is prohibited. Where single service cups (to be used but once) are supplied, both a sanitary container for unused cups and a receptacle for used cups shall be provided.

Occupational Noise Exposure/Hearing Protection

Feasible engineering or administrative controls shall be utilized to protect employees against sound levels in excess of those shown in the permissible noise exposure table show below. When engineering or administrative controls fail to reduce sound levels within the limits of the table below, ear protective devices shall be provided and used. In all cases where the sound levels exceed the values shown in the table below, a continuing, effective hearing conservation program shall be administered.

A hearing conservation program in construction should include the following elements:

- Monitoring employee noise exposures
- Using engineering, work practice and administrative controls, and personal protective equipment
- Fitting each overexposed employee with appropriate hearing protectors
- Training employees in the effects of noise and protection measures
- Explaining procedures for preventing further hearing loss

Exposure to impulsive or impact noise should not exceed 140 dB peak sound pressure level. Plain cotton is not an acceptable protective device. Types of hearing protection include ear plugs, canal caps, and ear muffs.

Duration Per Day, Hours	Sound Level dBA Slow Response
8	90
6	92
4	95
3	97
2	100
1½	102
1	105
½	110
¼ or less	115

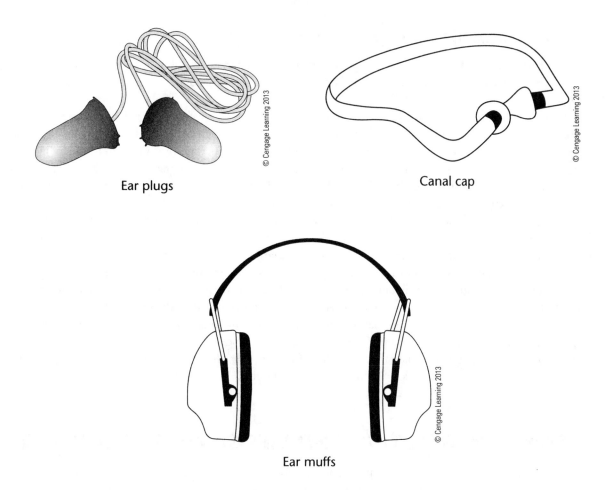

Ear plugs

Canal cap

Ear muffs

Illumination

Lighting should be adequate for a safe work area. Construction areas, ramps, runways, corridors, offices, shops, and storage areas shall be lighted to not less than the minimum illumination intensities shown below while any work is in progress:

Foot-candles	Area of Operation
5	General construction area lighting
3	General construction areas, concrete placement, excavation, waste areas, accessways, active storage areas, loading platforms, refueling, and field maintenance areas
5	Indoor warehouses, corridors, hallways, and exitways
5	Tunnels, shafts, and general underground work areas
10	General construction plant and shops
30	First-aid stations, infirmaries, and offices

Hazard Communication (HazCom)

The OSHA HazCom regulation for construction is taken directly from the General Industry standards: 1910.1200. They are relevant to the construction industry, because OSHA has incorporated the standard into the CFR Part 1926.

Employers shall develop, implement, and maintain at the workplace a written hazard communication program for their workplaces. Employers must inform their employees of the availability of the program, including the required list(s) of hazardous chemicals, and material safety data sheets required. The employer shall ensure that each container of hazardous chemicals in the workplace is labeled, tagged, or marked with the identity of the hazardous chemical(s) contained therein and must show hazard warnings appropriate for employee protection.

Chemical manufacturers and importers shall obtain or develop a material safety data sheet for each hazardous chemical they produce or import. Employers shall have a material safety data sheet for each hazardous chemical they use.

Employers shall provide employees with information and training on hazardous chemicals in their work area at the time of their initial assignment, and whenever a new hazard is introduced into their work area. Employers shall also provide employees with information on any operations in their work area where hazardous chemicals are present, and the location and availability of the written hazard communication program, including the required list(s) of hazardous chemicals, and material safety data sheets required by the standard.

Employers who produce, use, or store hazardous chemicals at multiemployer workplaces shall additionally ensure that their hazard communication program includes the methods the employer will use to provide other employer(s) with a copy of the material safety data sheet for hazardous chemicals. Further information on complying with the Hazard Communication Standard can be found in Resource C located at the back of this book and at dewalt.com/guides.

PERSONAL PROTECTIVE AND LIFE-SAVING EQUIPMENT

Protective equipment, including personal protective equipment (PPE) for eyes, face, head, and extremities, protective clothing, respiratory devices, and protective shields and barriers, shall be provided, used, and maintained in a sanitary and reliable condition wherever it is necessary by reason of hazards of processes or environment, chemical hazards, radiological hazards, or mechanical irritants encountered in a manner capable of causing injury or impairment in the function of any part of the body through absorption, inhalation, or physical contact.

Employer Payment for PPE

OSHA issued a final rule on Employer Payment for Personal Protective Equipment in November 2007. The rule requires employers in many industries, including construction, to provide most types of required PPE at no cost to the worker.

Employee-Owned Equipment

Where employees provide their own protective equipment, the employer shall be responsible to assure its adequacy, including proper maintenance, and sanitation of such equipment.

Foot Protection

Construction companies should evaluate their workplace to determine the need for foot protection. A standard practice in construction is requiring workers to wear safety-toe boots. If this is the jobsite requirement, safety-toe footwear for employees shall meet the requirements and specifications in American National Standard for Men's Safety-Toe Footwear, ANSI Z41.1-1967.

Head Protection

Employees working in areas where there is a possible danger of head injury from impact, or from falling or flying objects, or from electrical shock and burns shall be protected by protective hard hats. Hard hats for the protection of employees against impact and penetration of falling and flying objects shall meet the specifications contained in American National Standards Institute, ANSI Z89.1-1969, Safety Requirements for Industrial Head Protection. Hard hats for the head protection of employees exposed to high voltage electrical shock and burns shall meet the specifications contained in American National Standards Institute, ANSI Z89.2-1971.

Hard hats are a symbol of safety and should be a standard part of all workers' uniforms. Although OSHA only requires hard hats when there is a possible danger of head injury, it is recommended that construction companies take this a step further and require hard hats at all times on their construction sites.

Ear Protection

Refer to the occupation noise exposure/hearing protection paragraph earlier in this chapter.

Eye and Face Protection

Employees shall be provided with eye and face protection equipment when machines or operations present potential eye or face injury from physical, chemical, or radiation agents. Eye and face protection equipment shall meet the requirements specified in American National Standards Institute, ANSI Z87.1-1968, Practice for Occupational and Educational Eye and Face Protection.

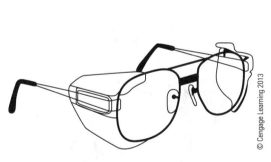

Prescription safety glasses with side shields

Standard safety glasses

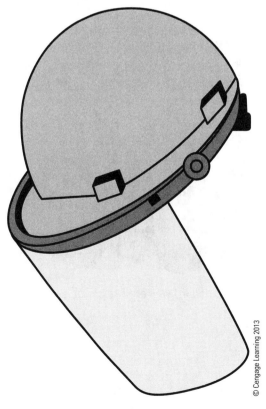

Face shield

Employees whose vision requires the use of prescription eyewear shall be protected by goggles or spectacles of one of the following types:

- Prescription eyewear with protective lenses rated for ANSI Z87.1 with side shields
- Safety glasses or goggles that can be worn over prescription eyewear without disturbing the adjustment of the eyewear
- Safety glasses or goggles that incorporate corrective lenses mounted behind the protective lenses

Face and eye protection equipment shall be kept clean and in good repair. The use of this type of equipment with structural or optical defects shall be prohibited.

Eye and face protectors shall meet the following minimum requirements:

- Provide adequate protection against the particular hazards for which they are designed
- Be reasonably comfortable when worn under the designated conditions
- Fit snugly and not unduly interfere with the movements of the wearer
- Be durable
- Be capable of being disinfected
- Be easily cleanable

Respiratory Protection

Respiratory hazards are present in many jobs and workplaces. There's a risk of serious illness or even death from inhaling dangerous airborne substances or working where the air contains too little oxygen.

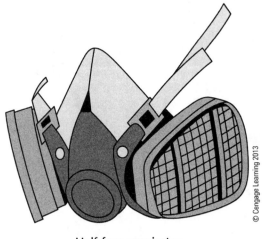

Half-face respirator

© Cengage Learning 2013

Full-face respirator

© Cengage Learning 2013

Supplied-air respirator

OSHA has always tried to protect against airborne hazards by requiring respirator use when ventilation and other controls can't make the air safe to breathe. A respirator is simply an enclosure that covers the nose and mouth or sometimes the entire face and head.

Employers need to determine whether there's a risk to employee health:

- Evaluate the workplace's respiratory hazards
- Identify relevant workplace and user factors that could affect respiratory hazards and health
- Develop a written respiratory protection program and appoint a qualified administrator
- Select and provide respirators appropriate for the hazards and certified by the National Institute for Occupational Safety and Health (NIOSH)

If workers are designated to use respiratory protection, they must receive:

- Training on the written respiratory protection program
- Medical evaluation—to determine the worker is healthy enough to wear a respirator
- Fit testing—to determine that the respirator chosen properly fits the worker

Working Over or Near Water

Employees working over or near water, where the danger of drowning exists, shall be provided with U.S. Coast Guard-approved life jacket or buoyant work vests. Prior to and after each use, the buoyant work vests or life preservers shall be inspected for defects that would alter their strength or buoyancy. Defective units shall not be used. Ring buoys with at least 90 feet of line shall be provided and readily available for emergency rescue operations. The distance between ring buoys shall not exceed 200 feet. At least one lifesaving skiff shall be immediately available at locations where employees are working over or adjacent to water.

FIRE PROTECTION AND PREVENTION

The employer shall be responsible for the development of a fire protection program to be followed throughout all phases of the construction and demolition work, and shall provide for the firefighting equipment as specified by OSHA. As fire hazards occur, there shall be no delay in providing the necessary equipment. Access to all available firefighting equipment shall be maintained at all times. All firefighting equipment, provided by the employer, shall be conspicuously located. All firefighting equipment shall be periodically inspected and maintained in operating condition. Defective equipment shall be immediately replaced.

Portable Firefighting Equipment

A fire extinguisher, rated not less than 2A, shall be provided for each 3,000 square feet of the protected building area, or major fraction thereof. Travel distance from any point of the protected area to the nearest fire extinguisher shall not exceed 100 feet. One 55-gallon open drum of water with two fire pails may be substituted for a fire extinguisher having a 2A rating. A ½-inch diameter garden-type hose line, not to exceed 100 feet in length and equipped with a nozzle, may be substituted for a 2A-rated fire extinguisher, providing it is capable of discharging a minimum of 5 gallons per minute with a minimum hose stream range of 30 feet horizontally. The garden-type hose lines shall be mounted on conventional racks or reels. The number and location of hose racks or reels shall be such that at least one hose stream can be applied to all points in the area.

One or more fire extinguishers, rated not less than 2A, shall be provided on each floor. In multistory buildings, at least one fire extinguisher shall be located adjacent to stairways. Extinguishers and water drums, subject to freezing, shall be protected from freezing.

A fire extinguisher, rated not less than 10B, shall be provided within 50 feet of wherever more than 5 gallons of flammable or combustible liquids or 5 pounds of flammable gas are being used on the jobsite. This requirement does not apply to the integral fuel tanks of motor vehicles. Carbon tetrachloride and other toxic vaporizing liquid fire extinguishers are prohibited.

Portable fire extinguishers shall be inspected periodically and maintained in accordance with Maintenance and Use of Portable Fire Extinguishers, NFPA No. 10A-1970.

Where the employer has provided portable fire extinguishers for employee use in the workplace, the employer shall also provide training to familiarize employees with the general principles of fire extinguisher use and the hazards involved with incipient stage fire fighting. This training shall be upon initial employment and at least annually thereafter.

The employer shall assure that portable fire extinguishers are maintained in a fully charged and operable condition and kept in their designated places at all times except during use. The employer shall assure that portable fire extinguishers are subjected to an annual maintenance check. Stored pressure extinguishers do not require an internal examination. The employer shall record the annual maintenance date and retain this record for one year after the last entry or the life of the shell, whichever is less.

Fire Prevention

Electrical wiring and equipment for light, heat, or power purposes shall be installed in compliance with the requirements of OSHA Regulation, Subpart K, Electrical.

Internal combustion engine powered equipment shall be so located that the exhausts are well away from combustible materials. When the exhausts are piped to outside the building under construction, a clearance of at least 6 inches shall be maintained between such piping and combustible material.

Smoking shall be prohibited at or in the vicinity of operations which constitute a fire hazard, and shall be conspicuously posted: "No Smoking or Open Flame."

Portable battery powered lighting equipment, used in connection with the storage, handling, or use of flammable gases or liquids shall be of the type approved for the hazardous locations.

The nozzle of air, inert gas, and steam lines or hoses, when used in the cleaning or ventilation of tanks and vessels that contain hazardous concentrations of flammable gases or vapors, shall be bonded to the tank or vessel shell. Bonding devices shall not be attached or detached in hazardous concentrations of flammable gases or vapors.

Temporary Heating Devices

Fresh air shall be supplied in sufficient quantities to maintain the health and safety of workers. Where a natural means of fresh air supply is inadequate, mechanical ventilation shall be provided. When heaters are used in confined spaces, special care shall be taken to provide sufficient ventilation in order to ensure proper combustion, maintain the health and safety of workers, and limit temperature rise in the area.

Heaters not suitable for use on wood floors shall not be set directly upon them or other combustible materials. When such heaters are used, they shall rest on suitable heat insulating material or at least 1-inch concrete, or equivalent. The insulating material shall extend beyond the heater 2 feet or more in all directions.

Heaters used in the vicinity of combustible tarpaulins, canvas, or similar coverings shall be located at least 10 feet from the coverings. The coverings shall be securely fastened to prevent ignition or upsetting of the heater due to wind action on the covering or other material.

Solid fuel salamanders are prohibited in buildings and on scaffolds.

Classification of Portable Fire Extinguishers

Portable fire extinguishers are classified to indicate their ability to handle specific classes and sizes of fires. Labels on extinguishers indicate the class and relative size of fire that they can be expected to handle.

- Class A extinguishers are used on fires involving ordinary combustibles, such as wood, cloth, and paper. An extinguisher for Class A fires could have any one of the following ratings: 1-A, 2-A, 3-A, 4-A, 6-A, 10-A, 20-A, 30-A, and 40-A. A 4-A extinguisher, for example, should extinguish about twice as much fire as a 2-A extinguisher.

- Class B extinguishers are used on fires involving liquids, greases, and gases. An extinguisher for Class B fires could have any one of the following ratings: 1-B, 2-B, 5-B, 10-B, 20-B, 30-B, 40-B, and up to 640-B.

- Class C extinguishers are used on fires involving energized electrical equipment. Extinguishers rated for Class C fires are tested only for electrical conductivity. However, no extinguisher gets a Class C rating without a Class A and/or Class B rating.

- Class D extinguishers are used on fires involving metals such as magnesium, titanium, zirconium, sodium, and potassium. Class D extinguishers are tested on metal fires. The agent used depends on the metal for which the extinguisher was designed. Check the extinguisher faceplate for the unit's effectiveness on specific metals.

Fire extinguisher with visible label

HIGHWAY WORK ZONES

Being struck by vehicles or mobile equipment leads to many employee work zone fatalities or injuries. Work zones need traffic controls identified by signs, cones, barrels and barriers.

Drivers, employees on foot, and pedestrians must be able to see and understand the proper routes. Construction project managers determine traffic control plans within construction/demolition worksites.

Traffic control devices, signals, and message boards instruct drivers to follow paths away from where work is being done.

Approved traffic control devices, including cones, barrels, barricades, and delineator posts, are also used inside work zones.

- **Work Zone Protections:** Various concrete, water, sand, collapsible barriers, crash cushions, and truck-mounted attenuators can help limit motorist intrusions into construction work zones.

- **Flagging:** Flaggers should wear high visibility clothing with a fluorescent background and made of retro-reflective material. This makes employees visible for at least 1,000 feet in any direction. Check the label or packaging to ensure that the garments are performance class 2 or 3. Drivers should be warned with signs that there will be flaggers ahead. Flaggers should use STOP/SLOW paddles, paddles with lights, or flags (only in emergencies).

- **Lighting:** Flagger stations should be illuminated. Lighting for employees on foot and for equipment operators should be at least 5 foot-candles or greater. Where available lighting is not sufficient, flares or chemical lighting should be used. Glare should be controlled or eliminated.

- **Training:** Flaggers must be trained/certified and use authorized signaling methods.

- **Driving:** Seat belts and rollover protection should be used on equipment and vehicles as the manufacturer recommends.

OSHA has incorporated by reference the Manual on Uniform Traffic Control Devices (MUTCD). By incorporating this manual by reference, the manual becomes an enforceable regulation. For a copy of the MUTCD, visit this website: http://mutcd.fhwa.dot.gov/.

MATERIALS HANDLING, STORAGE, USE, AND DISPOSAL

Handling and storing materials involves diverse operations such as hoisting tons of steel with a crane, driving a truck loaded with concrete blocks, manually carrying bags and material, and stacking drums, barrels, kegs, lumber, or loose bricks.

The efficient handling and storing of materials is vital to the construction industry. These operations provide a continuous flow of raw materials, parts, and assemblies through the jobsite and ensure that materials are available when needed. Yet, the improper handling and storing of materials can cause costly injuries. Workers should remember to:

- Keep the back straight, bend at the knees, and stand with feet apart
- Hold the object close to the body with a good grip
- Carry smaller loads and get help if the load is too large
- Turn using feet instead of twisting the back

Workers frequently cite the weight and bulkiness of objects being lifted as major contributing factors to their injuries. The second factor frequently cited by workers as contributing to their injuries is body movement. Bending, followed by twisting and turning, is the more commonly cited movement that caused back injuries.

In addition, workers can be injured by falling objects, improperly stacked materials, or by various types of equipment. When manually moving materials, however, workers should be aware of potential injuries, including the following:

- Strains and sprains from improperly lifting loads or from carrying loads that are either too large or too heavy
- Fractures and bruises caused by being struck by materials, or by being caught in pinch points
- Cuts and bruises caused by falling materials that have been improperly stored or by incorrectly cutting ties or other securing devices

Since numerous injuries can result from improperly handling and storing materials, it is important to be aware of accidents that may occur from unsafe or improperly handled equipment and improper work practices and to recognize the methods for eliminating, or at least minimizing, the occurrence of those accidents. Consequently, employers and employees can and should examine their workplaces to detect any unsafe or unhealthful conditions, practices, or equipment and take the necessary steps to correct them.

Methods of Prevention

General safety principles can help reduce workplace accidents. These include work practices, ergonomic principles, and training and education. Whether moving materials manually or mechanically, employees should be aware of the potential hazards associated with the task at hand and know how to exercise control over their workplaces to minimize the danger.

Moving, Handling, and Storing Materials

When manually moving materials, employees should seek help when a load is so bulky it cannot be properly grasped or lifted, when they cannot see around or over it, or when a load cannot be safely handled.

When an employee is placing blocks under raised loads, the employee should ensure that the load is not released until his or her hands are clearly removed from the load. Blocking materials and timbers should be large and strong enough to support the load safely. Materials with evidence of cracks, rounded corners, splintered pieces, or dry rot should not be used for blocking.

Handles and holders should be attached to loads to reduce the chances of getting fingers pinched or smashed. Workers also should use appropriate protective equipment. For loads with sharp or rough edges, wear gloves or other hand and forearm protection. To avoid injuries to the hands and eyes, use gloves and eye protection. When the loads are heavy or bulky, the mover should also wear steel-toed safety shoes or boots to prevent foot injuries if the worker slips or accidentally drops a load.

When mechanically moving materials, avoid overloading the equipment by letting the weight, size, and shape of the material being moved dictate the type of equipment used for transporting it. All materials handling equipment has rated capacities that determine the maximum weight the equipment can safely handle and the conditions under which it can handle those weights. The equipment-rated capacities must be displayed on each piece of equipment and must not be exceeded except for load testing.

When picking up items with a powered industrial truck, the load must be centered on the forks and as close to the mast as possible to minimize the potential for the truck tipping or the load falling. A lift truck must never be overloaded because it would be hard to control and could easily tip over. Extra weight must not be placed on the rear of a counterbalanced forklift to offset an overload. The load must be at the lowest position for traveling, and the truck manufacturer's operational requirements must be followed. All stacked loads must be correctly piled and cross-tiered, where possible. Precautions also should be taken when stacking and storing material.

Stored materials must not create a hazard. Storage areas must be kept free from accumulated materials that may cause tripping, fires, or explosions, or that may contribute to the harboring of rats and other pests. When stacking and piling materials, it is important to be aware of such factors as the materials' height and weight, how accessible the stored materials are to the user, and the condition of the containers where the materials are being stored.

When stacking materials, height limitations should be observed. For example, lumber must be stacked no more than 16 feet high if it is handled manually; 20 feet is the maximum stacking height if a forklift is used. For quick reference, walls or posts may be painted with stripes to indicate maximum stacking heights.

Used lumber must have all nails removed before stacking. Lumber must be stacked and leveled on solidly supported bracing. The stacks must be stable and self-supporting. Stacks of loose bricks should not be more than 7 feet in height. When these stacks reach a height of 4 feet, they should be tapered back 2 inches for every foot of height above the 4-foot level. When masonry blocks are stacked higher than 6 feet, the stacks should be tapered back one-half block for each tier above the 6-foot level.

Bags and bundles must be stacked in interlocking rows to remain secure. Bagged material must be stacked by stepping back the layers and cross-keying the bags at least every ten layers. To remove bags from the stack, start from the top row first. Baled paper and rags stored inside a building must not be closer than 18 inches to the walls, partitions, or sprinkler heads. Boxed materials must be banded or held in place using cross-ties or shrink plastic fiber.

Drums, barrels, and kegs must be stacked symmetrically. If stored on their sides, the bottom tiers must be blocked to keep them from rolling. When stacked on end, put planks, sheets of plywood dunnage, or pallets between each tier to make a firm, flat, stacking surface. When stacking materials two or more tiers high, the bottom tier must be chocked on each side to prevent shifting in either direction.

When stacking, consider the need for availability of the material. Material that cannot be stacked due to size, shape, or fragility can be safely stored on shelves or in bins. Structural steel, bar stock, poles, and other cylindrical materials, unless in racks, must be stacked and blocked to prevent spreading or tilting. Pipes and bars should not be stored in racks that face main aisles; this could create a hazard to passers-by when supplies are being removed.

Using Materials Handling Equipment

To reduce potential accidents associated with workplace equipment, employees need to be trained in the proper use and limitations of the equipment they operate. This includes knowing how to effectively use equipment such as conveyors, powered industrial trucks, and slings.

Conveyors

When using conveyors, workers' hands may be caught in nip points where the conveyor runs over support members or rollers; workers may be struck by material falling off the conveyor; or they may become caught on or in the conveyor, thereby being drawn into the conveyor path.

To reduce the severity of an injury, an emergency button or pull cord designed to stop the conveyor must be installed at the employee's work station. Continuously accessible conveyor belts should have an emergency stop cable that extends the entire length of the conveyor belt so that the cable can be accessed from any location along the belt. The emergency stop switch must be designed to be reset before the conveyor can be restarted. Before restarting a conveyor that has stopped due to an overload, appropriate personnel must inspect the conveyor and clear the stoppage before restarting. Employees must never ride on a materials handling conveyor. Where a conveyor passes over work areas or aisles, guards must be provided to keep employees from being struck by falling material. If the crossover is low enough for workers to run into, it must be guarded to protect employees and either marked with a warning sign or painted a bright color.

Screw conveyors must be completely covered except at loading and discharging points. At those points, guards must protect employees against contacting the moving screw; the guards are movable, and they must be interlocked to prevent conveyor movement when not in place.

Powered Industrial Trucks

Workers who must handle and store materials often use fork trucks, platform lift trucks, motorized hand trucks, and other specialized industrial trucks powered by electrical motors or internal combustion engines. Affected workers, therefore, should be aware of the safety requirements pertaining to fire protection, and the design, maintenance, and use of these trucks.

Workers who will operate powered industrial trucks must be trained and certified to operate the truck. The operator must be trained by a qualified trainer and the training should include classroom and field activities, including a road test. The training must be specific to the type of powered industrial truck and is only good for three years.

Safety precautions the user can observe when operating or maintaining powered industrial trucks include:

- Fitting high-lift rider trucks with an overhead guard, unless operating conditions do not permit
- Equipping fork trucks with a vertical load backrest extension according to manufacturers' specifications, if the load presents a hazard
- Locating battery charging installations in areas designated for that purpose
- Providing facilities for flushing and neutralizing spilled electrolytes when changing or recharging a battery to prevent fires, to protect the charging apparatus from being damaged by the trucks, and to adequately ventilate fumes in the charging area from gassing batteries
- Providing conveyor, overhead hoist, or equivalent materials handling equipment for handling batteries
- Installing auxiliary directional lighting on the truck where general lighting is less than two lumens per square foot
- Avoiding placing arms and legs between the uprights of the mast or outside the running lines of the truck
- Setting brakes and putting wheel blocks or other adequate protection in place to prevent movement of trucks, trailers, or railroad cars when using trucks to load or unload materials onto train boxcars
- Allowing sufficient headroom under overhead installations, lights, pipes, and sprinkler systems
- Providing personnel on the loading platform the means to shut off power to the truck
- Properly securing dock boards or bridge plates so they won't move when equipment moves over them
- Handling only stable or safely arranged loads, and exercising caution when handling loads
- Disconnecting the battery of trucks whose electrical systems are in need of repair prior to such repairs
- Ensuring that replacement parts of any industrial truck be equivalent in safety to the original ones

Modifications and additions that affect capacity and safe operation of the trucks shall not be performed by an owner or user without the manufacturer's prior written approval. In these cases, capacity, operation, and maintenance instruction plates and tags or decals must be changed to reflect the new information. If the truck is equipped with front-end attachments that are not factory installed, the user should request that the truck be marked to identify these attachments and show the truck's approximate weight, including the installed attachment, when it is at maximum elevation with its load laterally centered.

Sling Safety

There are many safe work practices to follow while using slings for materials handling:

1. Workers should accept the responsibility for their own actions and become competent and careful employees. Their own life and the life of their fellow workers or others may depend on it.

2. Workers should learn as much as they can about the materials with which they will be working. Slings come in many different types, one of which is right for the job.

3. Workers should analyze the load to be moved—in terms of size, weight, shape, temperature, and sensitivity—then they should choose the sling which best meets those needs.

4. Workers should always inspect all the equipment before and after a move. Always be sure to give equipment whatever "in-service" maintenance it may need.

5. Workers should use safe lifting practices. Every worker should use the proper lifting technique for the type of sling and the type of load.

TOOLS—HAND AND POWER

Tools are such a common part of our lives that it is difficult to remember that they may pose hazards. All tools are manufactured with safety in mind but, tragically, a serious accident often occurs before steps are taken to search out and avoid or eliminate tool-related hazards.

In the process of removing or avoiding the hazards, workers must learn to recognize the hazards associated with the different types of tools and the safety precautions necessary to prevent those hazards.

Hand Tools

Employees who use hand tools and who are exposed to the hazards of falling, flying, abrasive and splashing objects, or exposed to harmful dusts, fumes, mists, vapors, or gases must be provided with the particular personal equipment necessary to protect them from the hazard.

All hazards involved in the use of hand tools can be prevented by following five basic safety rules:

- Keep all tools in good condition with regular maintenance.
- Use the right tool for the job.
- Examine each tool for damage before use.
- Operate according to the manufacturer's instructions.
- Provide and use the proper protective equipment.

Hand tools are nonpowered. They include anything from axes to wrenches. The greatest hazards posed by hand tools result from misuse and improper maintenance. Some examples:

- Using a screwdriver as a chisel may cause the tip of the screwdriver to break and fly, hitting the user or other employees.
- If a wooden handle on a tool such as a hammer or an axe is loose, splintered, or cracked, the head of the tool may fly off and strike the user or another worker.
- A wrench must not be used if its jaws are sprung, because it might slip.
- Impact tools such as chisels, wedges, or drift pins are unsafe if they have mushroomed heads. The heads might shatter on impact, sending sharp fragments flying.

The employer is responsible for the safe condition of tools and equipment used by employees, but the employees have the responsibility for properly using and maintaining tools.

Employers should caution employees that saw blades, knives, or other tools be directed away from aisle areas and other employees working in close proximity. Knives and scissors must be sharp. Dull tools can be more hazardous than sharp ones.

Appropriate personal protective equipment, such as safety glasses and gloves should be worn due to hazards that may be encountered while using portable power tools and hand tools.

Safety requires that floors be kept as clean and dry as possible to prevent accidental slips with or around dangerous hand tools.

Around flammable substances, sparks produced by iron and steel hand tools can be a dangerous ignition source. Where this hazard exists, spark-resistant tools made from brass, plastic, aluminum, or wood will provide for safety.

Power Tools

Employees who use power tools and who are exposed to the hazards of falling, flying, abrasive and splashing objects, or exposed to harmful dusts, fumes, mists, vapors, or gases must be provided with the particular personal equipment necessary to protect them from the hazard.

All hazards involved in the use of power tools can be prevented by following five basic safety rules:

- Keep all tools in good condition with regular maintenance.
- Use the right tool for the job.
- Examine each tool for damage before use.
- Operate according to the manufacturer's instructions.
- Provide and use the proper protective equipment.

Power tools can be hazardous when improperly used. There are several types of power tools, based on the power source they use: electric, pneumatic, liquid fuel, hydraulic, and powder-actuated.

Employees should be trained in the use of all tools—not just power tools. They should understand the potential hazards as well as the safety precautions to prevent those hazards from occurring. The following general precautions should be observed by a power tool user:

- Never carry a tool by the cord or hose.
- Never yank the cord or the hose to disconnect it from the receptacle.
- Keep cords and hoses away from heat, oil, and sharp edges.
- Disconnect tools when not in use, before servicing, and when changing accessories such as blades, bits, and cutters.
- All observers should be kept at a safe distance away from the work area.
- Secure work with clamps or a vise, freeing both hands to operate the tool.
- Avoid accidental starting. The worker should not hold a finger on the switch button while carrying a plugged-in tool.
- Tools should be maintained with care. They should be kept sharp and clean for the best performance. Follow instructions in the user's manual for lubricating and changing accessories.
- Be sure to keep good footing and maintain good balance.
- The proper apparel should be worn. Loose clothing, ties, or jewelry can become caught in moving parts.
- All portable electric tools that are damaged shall be removed from use and tagged "Do Not Use."

Machine Guarding

Hazardous moving parts of a power tool need to be safeguarded. For example, belts, gears, shafts, pulleys, sprockets, spindles, drums, fly wheels, chains, or other reciprocating, rotating, or moving parts of equipment must be guarded if such parts are exposed to contact by employees. Guards, as necessary, should be provided to protect the operator and others from the following:

- Point of operation
- In-running nip points
- Rotating parts
- Flying chips and sparks

Safety guards must never be removed when a tool is being used. For example, portable circular saws must be equipped with guards. An upper guard must cover the entire blade of the saw. A retractable lower guard must cover the teeth of the saw, except when it makes contact with the work material. The lower guard must automatically return to the covering position when the tool is withdrawn from the work.

Courtesy of Stanley Black & Decker

Saw with proper guard

Safety Switches

The following hand-held powered tools must be equipped with a momentary contact "on-off" control switch: drills; tappers; fastener drivers; horizontal, vertical and angle grinders with wheels larger than 2 inches in diameter; disc and belt sanders; reciprocating saws; saber saws; and other similar tools. These tools also may be equipped with a lock-on control provided that turnoff can be accomplished by a single motion of the same finger or fingers that turn it on.

The following hand-held powered tools may be equipped with only a positive "on-off" control switch: platen sanders, disc sanders with discs 2 inches or less in diameter; grinders with wheels 2 inches or less in diameter; routers, planers, laminate trimmers, nibblers, shears, scroll saws and jigsaws with blade shanks ¼-inch wide or less.

Other hand-held power tools such as circular saws having a blade diameter greater than 2 inches, chain saws, and percussion tools without positive accessory holding means must be equipped with a constant pressure switch that will shut off the power when the pressure is released.

Electric Tools

Employees using electric tools must be aware of several dangers; the most serious is the possibility of electrocution. Among the chief hazards of electric-powered tools are burns and slight shocks, which can lead to injuries or even heart failure. Under certain conditions, even a small amount of current can result in fibrillation of the heart and eventual death. A shock also can cause the user to fall off a ladder or other elevated work surface.

To protect the user from shock, tools must either have a three-wire cord with ground and be grounded, be double insulated, or be powered by a low-voltage isolation transformer. Three-wire cords contain two current-carrying conductors and a grounding conductor. One end of the grounding conductor connects to the tool's metal housing. The other end is grounded through a prong on the plug. Anytime an adapter is used to accommodate a two-hole receptacle, the adapter wire must be attached to a known ground. The third prong should never be removed from the plug.

Double insulation is more convenient. The user and the tools are protected in two ways: by normal insulation on the wires inside, and by a housing that cannot conduct electricity to the operator in the event of a malfunction.

These general practices should be followed when using electric tools:

- Electric tools should be operated within their design limitations.
- Gloves and safety footwear are recommended during use of electric tools.
- When not in use, tools should be stored in a dry place.
- Electric tools should not be used in damp or wet locations.
- Work areas should be well lighted.

Powered Abrasive Wheel Tools

Powered abrasive grinding, cutting, polishing, and wire buffing wheels create special safety problems because they may throw off flying fragments.

Courtesy of Stanley Black & Decker

Angle grinder with proper guard

Before an abrasive wheel is mounted, it should be inspected closely and sound-or-ring tested to be sure that it is free from cracks or defects. To test, wheels should be tapped gently with a light, nonmetallic instrument. If they sound cracked or dead, they could fly apart in operation and so must not be used. A sound and undamaged wheel will give a clear metallic tone or "ring."

To prevent the wheel from cracking, the user should be sure it fits freely on the spindle. The spindle nut must be tightened enough to hold the wheel in place, without distorting the flange. Follow the manufacturer's recommendations. Care must be taken to assure that the spindle wheel will not exceed the abrasive wheel specifications.

Due to the possibility of a wheel disintegrating (exploding) during start-up, the employee should never stand directly in front of the wheel as it accelerates to full operating speed.

Portable grinding tools need to be equipped with safety guards to protect workers not only from the moving wheel surface, but also from flying fragments in case of breakage. In addition, when using a powered grinder:

- Always use eye protection.
- Turn off the power when not in use.
- Never clamp a handheld grinder in a vise.

Pneumatic Tools

Pneumatic tools are powered by compressed air and include chippers, drills, hammers, and sanders. There are several dangers encountered in the use of pneumatic tools. The main one is the danger of getting hit by one of the tool's attachments or by some kind of fastener the worker is using with the tool.

Eye protection is required and face protection is recommended for employees working with pneumatic tools. Noise is another hazard. Working with noisy tools such as jackhammers requires proper, effective use of hearing protection.

When using pneumatic tools, employees must check to see that they are fastened securely to the hose to prevent them from becoming disconnected. A short wire or positive locking device attaching the air hose to the tool will serve as an added safeguard.

A safety clip or retainer must be installed to prevent attachments, such as chisels on a chipping hammer, from being unintentionally shot from the barrel.

Screens must be set up to protect nearby workers from being struck by flying fragments around chippers, riveting guns, staplers, or air drills.

Compressed air guns should never be pointed toward anyone. Users should never "dead-end" it against themselves or anyone else.

Powder-Actuated Tools

Powder-actuated tools operate like a loaded gun and should be treated with the same respect and precautions. In fact, they are so dangerous that they must be operated only by specially trained employees.

Safety precautions to remember include the following:

- These tools should not be used in an explosive or flammable atmosphere.
- Before using the tool, the worker should inspect it to determine that it is clean, that all moving parts operate freely, and that the barrel is free from obstructions.
- The tool should never be pointed at anybody.
- The tool should not be loaded unless it is to be used immediately. A loaded tool should not be left unattended, especially where it would be available to unauthorized persons.
- Hands should be kept clear of the barrel end. To prevent the tool from firing accidentally, two separate motions are required for firing: one to bring the tool into position, and another to pull the trigger. The tools must not be able to operate until they are pressed against the work surface with a force of at least 5 pounds greater than the total weight of the tool.

If a powder-actuated tool misfires, the employee should wait at least 30 seconds, then try firing it again. If it still will not fire, the user should wait another 30 seconds so that the faulty cartridge is less likely to explode, then carefully remove the load. The bad cartridge should be put in water.

Suitable eye and face protection are essential when using a powder-actuated tool.

The muzzle end of the tool must have a protective shield or guard centered perpendicularly on the barrel to confine any flying fragments or particles that might otherwise create a hazard when the tool is fired. The tool must be designed so that it will not fire unless it has this kind of safety device.

All powder-actuated tools must be designed for varying powder charges so that the user can select a powder level necessary to do the work without excessive force.

If the tool develops a defect during use it should be tagged and taken out of service immediately until it is properly repaired.

Fasteners

When using powder-actuated tools to apply fasteners, there are some precautions to consider. Fasteners must not be fired into material that would let them pass through to the other side. The fastener must not be driven into materials like brick or concrete any closer than 3 inches to an edge or corner. In steel, the fastener must not come any closer than ½ inch from a corner or edge. Fasteners must not be driven into very hard or brittle materials, which might chip or splatter or make the fastener ricochet.

An alignment guide must be used when shooting a fastener into an existing hole. A fastener must not be driven into a spalled area caused by an unsatisfactory fastening.

Hydraulic Power Tools

Hydraulic power tools present many unique safety hazards. The user of any hydraulic power tool should consult the manufacturers' user manual to make sure they understand all the characteristics of the particular tool they are using. Here are some general safety precautions for hydraulic power tools:

Do

- Do inspect the tool and all hoses and connections before use
- Do connect all hoses and hose couplers before energizing
- Do use certified non-conductive hoses when working near electrical conductors

Don't

- Do not operate any damaged, improperly adjusted or improperly assemble tool
- Do not carry tools by their hoses
- Do not clean or inspect the tools while the hydraulic power source is connected
- Do not use conductive hoses near any electrical conductors

When using a hydraulic power tool, the user must be extremely careful and aware of possible hydraulic leaks. A hydraulic leak or burst may cause oil to be injected into the body or cause other severe injury. To reduce the risk of hydraulic leaks or bursts:

- Do not exceed the specified flow and pressure for the tool
- Do not exceed rated working pressure of the hydraulic hose
- Inspect the tool each day prior to use for leaks; do not feel for leaks with your hands
- Check all hoses, couplers, and connections

The fluid used in hydraulic power tools must be an approved fire-resistant fluid and must retain its operating characteristics at the most extreme temperatures to which it will be exposed. The manufacturer's recommended safe operating pressure for hoses, valves, pipes, filters, and other fittings must not be exceeded.

Jacks

All jacks—lever and ratchet jacks, screw jacks, and hydraulic jacks—must have a device that stops them from jacking up too high. Also, the manufacturer's load limit must be permanently marked in a prominent place on the jack and should not be exceeded.

A jack should never be used to support a lifted load. Once the load has been lifted, it must immediately be blocked up.

Use wooden blocking under the base if necessary to make the jack level and secure. If the lift surface is metal, place a 1-inch-thick hardwood block or equivalent between it and the metal jack head to reduce the danger of slippage.

To set up a jack, make certain of the following:

- The base rests on a firm level surface.
- The jack is correctly centered.
- The jack head bears against a level surface.
- The lift force is applied evenly.

Proper maintenance of jacks is essential for safety. All jacks must be inspected before each use and lubricated regularly. If a jack is subjected to an abnormal load or shock, it should be thoroughly examined to make sure it has not been damaged.

Hydraulic jacks exposed to freezing temperatures must be filled with an adequate antifreeze liquid.

General Safety Precautions

Employees who use hand and power tools and who are exposed to the hazards of falling, flying, abrasive, and splashing objects or exposed to harmful dusts, fumes, mists, vapors, or gases must be provided with the particular personal equipment necessary to protect them from the hazard.

All hazard involved in the use of power tools can be prevented by following five basic safety rules:

- Keep all tools in good condition with regular maintenance.
- Use the right tool for the job.
- Examine each tool for damage before use.
- Operate according to the manufacturer's instructions.
- Provide and use the proper protective equipment.

Employees and employers have a responsibility to work together to establish safe working procedures. If a hazardous situation is encountered, it should be brought to the attention of the proper individual immediately.

WELDING AND CUTTING

Gas Welding and Cutting

Transporting, Moving, and Storing Compressed Gas Cylinders

- Valve protection caps shall be in place and secured.
- When cylinders are hoisted, they shall be secured on a cradle, sling board, or pallet. They shall not be hoisted or transported by means of magnets or choker slings.
- Cylinders shall be moved by tilting and rolling them on their bottom edges. They shall not be intentionally dropped, struck, or permitted to strike each other violently.
- When cylinders are transported by powered vehicles, they shall be secured in a vertical position.

- Valve protection caps shall not be used for lifting cylinders from one vertical position to another. Bars shall not be used under valves or valve protection caps to pry cylinders loose when frozen. Warm, not boiling, water shall be used to thaw cylinders loose.
- Unless cylinders are firmly secured on a special carrier intended for this purpose, regulators shall be removed and valve protection caps put in place before cylinders are moved.
- A suitable cylinder truck, chain, or other steadying device shall be used to keep cylinders from being knocked over while in use.
- When work is finished, when cylinders are empty, or when cylinders are moved at any time, the cylinder valve shall be closed.
- Compressed gas cylinders shall be secured in an upright position at all times, and if necessary, for short periods of time while cylinders are actually being hoisted or carried.
- Oxygen cylinders in storage shall be separated from fuel-gas cylinders or combustible materials (especially oil or grease), a minimum distance of 20 feet (6.1 m) or by a noncombustible barrier at least 5 feet (1.5 m) high having a fire-resistance rating of at least one-half hour.
- Inside of buildings, cylinders shall be stored in a well-protected, well-ventilated, dry location, at least 20 feet (6.1 m) from highly combustible materials such as oil or excelsior. Cylinders should be stored in definitely assigned places away from elevators, stairs, or gangways. Assigned storage places shall be located where cylinders will not be knocked over or damaged by passing or falling objects or subject to tampering.

Placing Cylinders

- Cylinders shall be kept far enough away from the actual welding or cutting operation so that sparks, hot slag, or flame will not reach them. When this is impractical, fire-resistant shields shall be provided.
- Cylinders shall be placed where they cannot become part of an electrical circuit. Electrodes shall not be struck against a cylinder to strike an arc.
- Fuel-gas cylinders shall be placed with valve end up whenever they are in use. They shall not be placed in a location where they would be subject to open flame, hot metal, or other sources of artificial heat.
- Cylinders containing oxygen or acetylene or other fuel gas shall not be taken into confined spaces.

Treatment of Cylinders

- Cylinders, whether full or empty, shall not be used as rollers or supports.
- No person other than the gas supplier shall attempt to mix gases in a cylinder. No one except the owner of the cylinder or person authorized by him shall refill a cylinder. No one shall use a cylinder's contents for purposes other than those intended by the supplier. All cylinders used shall meet the Department of Transportation requirements published in 49 CFR Part 178, Subpart C, Specification for Cylinders.
- No damaged or defective cylinder shall be used.

Use of Fuel Gas

The employer shall thoroughly instruct employees in the safe use of fuel gas, as follows:

- Fuel gas shall not be used from cylinders through torches or other devices that are equipped with shutoff valves without reducing the pressure through a suitable regulator attached to the cylinder valve or manifold.
- Before a regulator to a cylinder valve is connected, the valve shall be opened slightly and closed immediately. (This action is generally termed "cracking" and is intended to clear the valve of dust or dirt that might otherwise enter the regulator.) The person cracking the valve shall stand to one side of the outlet,

not in front of it. The valve of a fuel gas cylinder shall not be cracked where the gas would reach welding work, sparks, flame, or other possible sources of ignition.

- The cylinder valve shall always be opened slowly to prevent damage to the regulator. For quick closing, valves of fuel-gas cylinders shall not be opened more than 1 ½ turns. When a special wrench is required, it shall be left in position on the stem of the valve while the cylinder is in use so that the fuel-gas flow can be shut off quickly in case of an emergency. In the case of manifolded or coupled cylinders, at least one such wrench shall always be available for immediate use. Nothing shall be placed on top of a fuel-gas cylinder, when in use, which may damage the safety device or interfere with the quick closing of the valve.

- Before a regulator is removed from a cylinder valve, the cylinder valve shall always be closed and the gas released from the regulator.

- If, when the valve on a fuel-gas cylinder is opened, a leak is found around the valve stem, the valve shall be closed and the gland nut tightened. If this action does not stop the leak, the use of the cylinder shall be discontinued and it shall be properly tagged and removed from the work area. In the event that fuel gas should leak from the cylinder valve, rather than from the valve stem, and the gas cannot be shut off, the cylinder shall be properly tagged and removed from the work area. If a regulator attached to a cylinder valve will effectively stop a leak through the valve seat, the cylinder need not be removed from the work area.

- If a leak should develop at a fuse plug or other safety device, the cylinder shall be removed from the work area.

Fuel Gas and Oxygen Manifolds

Fuel-gas and oxygen manifolds shall bear the name of the substance they contain in letters at least 1-inch high, which shall be either painted on the manifold or on a sign permanently attached to it. These manifolds shall be placed in safe, well-ventilated, and accessible locations and not be located within enclosed spaces.

Manifold hose connections, including both ends of the supply hose that lead to the manifold, shall be such that the hose cannot be interchanged between fuel-gas and oxygen manifolds and supply header connections. Adapters shall not be used to permit the interchange of hose. Hose connections shall be kept free of grease and oil. When not in use, manifold and header hose connections shall be capped.

Nothing shall be placed on top of a manifold, when in use, which will damage the manifold or interfere with the quick closing of the valves.

Hose

Fuel-gas and oxygen hose shall be easily distinguishable from each other. The distinction may be made by different colors or by surface characteristics readily distinguishable by the sense of touch. Oxygen and fuel-gas hoses shall not be interchangeable. A single hose having more than one gas passage shall not be used.

When parallel sections of oxygen and fuel-gas hose are taped together, not more than 4 inches out of 12 inches shall be covered by tape.

All hose in use, carrying acetylene, oxygen, natural or manufactured fuel gas, or any gas or substance that may ignite or enter into combustion or be in any way harmful to employees, shall be inspected at the beginning of each working shift. Defective hose shall be removed from service.

Hose which has been subject to flashback, or which shows evidence of severe wear or damage, shall be tested to twice the normal pressure to which it is subject, but in no case less than 300 psi. Defective hose, or hose in doubtful condition, shall not be used.

Hose couplings shall be of the type that cannot be unlocked or disconnected by means of a straight pull without rotary motion.

Boxes used for the storage of gas hose shall be ventilated.

Hoses, cables, and other equipment shall be kept clear of passageways, ladders, and stairs.

Torches

Clogged torch tip openings shall be cleaned with suitable cleaning wires, drills, or other devices designed for such purpose.

Torches in use shall be inspected at the beginning of each working shift for leaking shutoff valves, hose couplings, and tip connections. Defective torches shall not be used.

Torches shall be lighted by friction lighters or other approved devices, and not by matches or from hot work.

Regulators and Gauges

Oxygen and fuel-gas pressure regulators, including their related gauges, shall be in proper working order while in use.

Oil and Grease Hazards

Oxygen cylinders and fittings shall be kept away from oil or grease. Cylinders, cylinder caps and valves, couplings, regulators, hose, and apparatus shall be kept free from oil or greasy substances and shall not be handled with oily hands or gloves. Oxygen shall not be directed at oily surfaces, greasy clothes, or within a fuel oil or other storage tank or vessel.

ARC WELDING AND CUTTING

Manual Electrode Holders

Only manual electrode holders that are specifically designed for arc welding and cutting, and are of a capacity capable of safely handling the maximum rated current required by the electrodes, shall be used.

Any current-carrying parts passing through the portion of the holder that the arc welder or cutter grips in his hand and the outer surfaces of the jaws of the holder shall be fully insulated against the maximum voltage encountered to ground.

Welding Cables and Connectors

All arc welding and cutting cables shall be of the completely insulated, flexible type, capable of handling the maximum current requirements of the work in progress, taking into account the duty cycle under which the arc welder or cutter is working.

Only cable free from repair or splices for a minimum distance of 10 feet from the cable end to which the electrode holder is connected shall be used, except that cables with standard insulated connectors or with splices whose insulating quality is equal to that of the cable are permitted.

Cables in need of repair shall not be used. When a cable, other than the cable lead referred to above, becomes worn to the extent of exposing bare conductors, the portion thus exposed shall be protected by means of rubber and friction tape or other equivalent insulation.

When it becomes necessary to connect or splice lengths of cable one to another, substantial insulated connectors of a capacity at least equivalent to that of the cable shall be used. If connections are effected by means of cable lugs, they shall be securely fastened together to give good electrical contact, and the exposed metal parts of the lugs shall be completely insulated.

Ground Returns and Machine Grounding

A ground-return cable shall have a safe current-carrying capacity equal to or exceeding the specified maximum output capacity of the arc welding or cutting unit that it services. When a single ground-return cable services more than one unit, its safe current-carrying capacity shall exceed the total specified maximum output capacities of all the units that it services.

Pipelines containing gases or flammable liquids, or conduits containing electrical circuits, shall not be used as a ground return.

When a structure or pipeline is employed as a ground-return circuit, it shall be determined that the required electrical contact exists at all joints. The generation of an arc, sparks, or heat at any point shall cause rejection of the structures as a ground circuit.

When a structure or pipeline is continuously employed as a ground-return circuit, all joints shall be bonded, and periodic inspections shall be conducted to ensure that no condition of electrolysis or fire hazard exists by virtue of such use.

The frames of all arc welding and cutting machines shall be grounded either through a third wire in the cable containing the circuit conductor or through a separate wire that is grounded at the source of the current. Grounding circuits, other than by means of the structure, shall be checked to ensure that the circuit between the ground and the grounded power conductor has resistance low enough to permit sufficient current to flow to cause the fuse or circuit breaker to interrupt the current.

All ground connections shall be inspected to ensure that they are mechanically strong and electrically adequate for the required current.

Operating Instructions

Employers shall instruct employees in the safe means of arc welding and cutting as follows:

- When electrode holders are to be left unattended, the electrodes shall be removed and the holders shall be so placed or protected that they cannot make electrical contact with employees or conducting objects.
- Hot electrode holders shall not be dipped in water; to do so may expose the arc welder or cutter to electric shock.
- When the arc welder or cutter has occasion to leave his work or to stop work for any appreciable length of time, or when the arc welding or cutting machine is to be moved, the power supply switch to the equipment shall be opened.
- Any faulty or defective equipment shall be reported to the supervisor.
- A disconnecting means shall be provided in the supply circuit for each motor-generated arc welder, and for each AC transformer and DC rectifier arc welder that is not equipped with a disconnect mounted as an integral part of the welder.
- A switch or circuit breaker shall be provided by which each resistance welder and its control equipment can be isolated from the supply circuit. The ampere rating of this disconnecting means shall not be less than the supply conductor ampacity.

Shielding

Whenever practicable, all arc welding and cutting operations shall be shielded by a noncombustible or flame-proof screen that will protect employees and other persons working in the vicinity from the direct rays of the arc.

Welding and Cutting Fire Protection

When practical, objects to be welded, cut, or heated shall be moved to a designated safe location or, if these objects cannot be readily moved, all movable fire hazards in the vicinity shall be taken to a safe place or otherwise protected. If these objects cannot be moved and if all the fire hazards cannot be removed, positive means shall be taken to confine the heat, sparks, and slag and to protect the immovable fire hazards from them.

No welding, cutting, or heating shall be done where the application of flammable paints, or the presence of other flammable compounds, or heavy dust concentrations create a hazard.

Suitable fire extinguishing equipment shall be immediately available in the work area and shall be maintained in a state or readiness for instant use.

When the welding, cutting, or heating operation is such that normal fire prevention precautions are not sufficient, additional personnel shall be assigned to guard against fire while the actual welding, cutting, or heating operation is being performed and for a sufficient period of time after completion of the work to ensure that no possibility of fire exists. Such personnel shall be instructed as to the specific anticipated fire hazards and how the firefighting equipment provided is to be used.

When welding, cutting, or heating is performed on walls, floors, and ceilings, since direct penetration of sparks or heat transfer may introduce a fire hazard to an adjacent area, the same precautions shall be taken on the opposite side as are taken on the side on which the welding is being performed.

For the elimination of possible fire in enclosed spaces as a result of gas escaping through leaking or improperly closed torch valves, the gas supply to the torch shall be positively shut off at some point outside the enclosed space whenever the torch is not to be used or whenever the torch is left unattended for a substantial period of time, such as during the lunch period. Overnight and at the change of shifts, the torch and hose shall be removed from the confined space. Open-end fuel-gas and oxygen hoses shall be immediately removed from enclosed spaces when they are disconnected from the torch or other gas-consuming device.

Except when the contents are being removed or transferred, drums, pails, and other containers that contain or have contained flammable liquids shall be kept closed. Empty containers shall be removed to a safe area apart from hot work operations or open flames.

Drums, containers, or hollow structures that have contained toxic or flammable substances shall, before welding, cutting, or heating is undertaken on them, either be filled with water or thoroughly cleaned of such substances and ventilated and tested.

Before heat is applied to a drum, container, or hollow structure, a vent or opening shall be provided for the release of any built-up pressure during the application of heat.

ELECTRICAL

Electricity has long been recognized as a serious workplace hazard, exposing employees to such dangers as electric shock, electrocution, fires, and explosions. The following regulations and safe work practices shall be used to prevent these serious workplace hazards.

Examination, Installation, and Use of Equipment

The employer must ensure that electrical equipment is free from recognized hazards that are likely to cause death or serious physical harm to employees. Safety of equipment must be determined by the following:

- Suitability for installation and use in conformity with the provisions of the standard. Suitability of equipment for an identified purpose may be evidenced by a listing, by labeling, or by certification for that identified purpose
- Mechanical strength and durability. For parts designed to enclose and protect other equipment, this includes the adequacy of the protection thus provided
- Electrical insulation
- Heating effects under conditions of use
- Arcing effects
- Classification by type, size, voltage, current capacity, and specific use
- Other factors that contribute to the practical safeguarding of employees who use or are likely to come in contact with the equipment

Guarding

Live parts of electric equipment operating at 50 volts or more must be guarded against accidental contact. Guarding of live parts must be accomplished as follows:

- Location in a cabinet, room, vault, or similar enclosure accessible only to qualified persons
- Use of permanent, substantial partitions or screens to exclude unqualified persons
- Location on a suitable balcony, gallery, or platform elevated and arranged to exclude unqualified persons
- Elevation of 8 feet or more above the floor

Entrance to rooms and other guarded locations containing exposed live parts must be marked with conspicuous warning signs forbidding unqualified persons to enter.

Electric installations that are over 600 volts and that are open to unqualified persons must be made with metal-enclosed equipment or enclosed in a vault or area controlled by a lock. In addition, equipment must be marked with appropriate caution signs.

Grounding of Equipment Connected by Cord and Plug

Exposed non-current-carrying metal parts of cord-and-plug connected equipment that may become energized must be grounded in the following situations:

- When in a hazardous (classified) location
- When operated at over 150 volts to ground, except for guarded motors and metal frames of electrically heated appliances if the appliance frames are permanently and effectively insulated from ground
- When one of the types of equipment listed below. But see Item 6 for exemption
  1. Handheld motor-operated tools
  2. Cord-and-plug connected equipment used in damp or wet locations or by employees standing on the ground or on metal floors or working inside metal tanks or boilers
  3. Portable and mobile X-ray and associated equipment
  4. Tools likely to be used in wet and/or conductive locations
  5. Portable hand lamps
  6. [Exemption] Tools likely to be used in wet and/or conductive locations need not be grounded if supplied through an isolating transformer with an ungrounded secondary of not over 50 volts. Listed or labeled portable tools and appliances protected by a system of double insulation, or its equivalent, need not be grounded. If such a system is employed, the equipment must be distinctively marked to indicate that the tool or appliance uses a system of double insulation

Safe Work Practices

Protection of Employees

The employer must not permit an employee to work near any part of an electric power circuit that the employee could contact in the course of work, unless the employee is protected against shock by de-energizing the circuit and grounding it or by guarding it effectively by insulation or other means.

Where the exact location of underground electric power lines is unknown, employees using jack hammers or hand tools that may contact a line must be provided with insulated protective gloves.

Even before work is begun, the employer must determine by inquiry, observation, or instruments where any part of an exposed or concealed energized electric power circuit is located. This is necessary because a person, tool, or machine could come into physical or electrical contact with the electric power circuit.

The employer is required to advise employees of the location of such lines, the hazards involved, and protective measures to be taken as well as to post and maintain proper warning signs.

Passageways and Open Spaces

The employer must provide barriers or other means of guarding to ensure that workspace for electrical equipment will not be used as a passageway during the time when energized parts of electrical equipment are exposed. Walkways and similar working spaces must be kept clear of electric cords. Other standards cover load ratings, fuses, cords, and cables.

Lockout and Tagging of Circuits

Tags must be placed on controls that are to be deactivated during the course of work on energized or de-energized equipment or circuits. Equipment or circuits that are de-energized must be rendered inoperative and have tags attached at all points where such equipment or circuits can be energized.

Safety-Related Maintenance and Environmental Considerations

Maintenance of Equipment

The employer must ensure that all wiring components and utilization equipment in hazardous locations are maintained in a dust-tight, dust-ignition-proof, or explosion-proof condition without loose or missing screws, gaskets, threaded connections, seals, or other impairments to a tight condition.

Environmental Deterioration of Equipment

Unless identified for use in the operating environment, no conductors or equipment can be located:

- In damp or wet locations
- Where exposed to gases, fumes, vapors, liquids, or other agents having a deteriorating effect on the conductors or equipment
- Where exposed to excessive temperatures

Control equipment, utilization equipment, and busways approved for use in dry locations only must be protected against damage from the weather during building construction.

For protection against corrosion, metal raceways, cable armor, boxes, cable sheathing, cabinets, elbows, couplings, fittings, supports, and support hardware must be of materials appropriate for the environment in which they are installed.

Ground-Fault Protection on Construction Sites

Insulation and grounding are two recognized means of preventing injury during electrical equipment operation. Conductor insulation may be provided by placing nonconductive material such as plastic around the conductor. Grounding may be achieved through the use of a direct connection to a known ground such as a metal cold water pipe.

Consider, for example, the metal housing or enclosure around a motor or the metal box in which electrical switches, circuit breakers, and controls are placed. Such enclosures protect the equipment from dirt and moisture and prevent accidental contact with exposed wiring. However, there is a hazard associated with housings and enclosures. A malfunction within the equipment—such as deteriorated insulation—may create an electrical shock hazard. Many metal enclosures are connected to a ground to eliminate the hazard. If a "hot" wire contacts a grounded enclosure, a ground fault results, which normally will trip a circuit breaker or blow a fuse. Metal enclosures and containers are usually grounded by connecting them with a wire going to ground. This wire is called an equipment grounding conductor. Most portable electric tools and appliances are grounded by this means. There is one disadvantage to grounding: a break in the grounding system may occur without the user's knowledge.

Insulation may be damaged by hard usage on the job or simply by aging. If this damage causes the conductors to become exposed, the hazards of shocks, burns, and fire will exist. Double insulation may be used as additional protection on the live parts of a tool, but double insulation does not provide protection against defective cords and plugs or against heavy moisture conditions.

The use of a ground-fault circuit interrupter (GFCI) is one method of overcoming grounding and insulation deficiencies.

What Is a GFCI?

The ground-fault circuit interrupter (GFCI) is a fast-acting circuit breaker that senses small imbalances in the circuit caused by current leakage to ground and, in a fraction of a second, shuts off the electricity. The GFCI continually matches the amount of current going to an electrical device against the amount of current returning from the device along the electrical path. Whenever the amount "going" differs from the amount "returning" by approximately 5 milliamps, the GFCI interrupts the electric power within as little as 1/40 of a second.

However, the GFCI will not protect the employee from line-to-line contact hazards (such as a person holding two "hot" wires or a hot and a neutral wire in each hand). It does provide protection against the most common form of electrical shock hazard—the ground fault. It also provides protection against fires, overheating, and destruction of insulation on wiring.

What Are the Hazards?

Because portable tools are widely used on construction sites, flexible cords often become necessary. Hazards are created when cords, cord connectors, receptacles, and cord-and-plug connected equipment are improperly used and maintained.

Generally, flexible cords are more vulnerable to damage than is fixed wiring. Flexible cords must be connected to devices and to fittings so as to prevent tension at joints and terminal screws. Because a cord is exposed, flexible, and unsecured, joints and terminals become more vulnerable. Flexible cord conductors are finely stranded for flexibility, but the strands of one conductor may loosen from under terminal screws and touch another conductor, especially if the cord is subjected to stress or strain.

A flexible cord may be damaged by activities on the job, by door or window edges, by staples or fastenings, by abrasion from adjacent materials, or simply by aging. If the electrical conductors become exposed, there is a danger of shocks, burns, or fire. A frequent hazard on a construction site is a cord assembly with improperly connected terminals.

When a cord connector is wet, hazardous leakage can occur to the equipment grounding conductor and to humans who pick up that connector if they also provide a path to ground. Such leakage is not limited to the face of the connector but also develops at any wetted portion of it.

When the leakage current of tools is below 1 ampere, and the grounding conductor has a low resistance, no shock should be perceived. However, should the resistance of the equipment grounding conductor increase, the current through the body will also increase. Thus, if the resistance of the equipment grounding conductor is significantly greater than 1 ohm, tools with even small leakages become hazardous.

Preventing and Eliminating Hazards

GFCIs can be used successfully to reduce electrical hazards on construction sites. Tripping of GFCIs—interruption of current flow—is sometimes caused by wet connectors and tools. It is good practice to limit exposure of connectors and tools to excessive moisture by using watertight or sealable connectors. Providing more GFCIs or shorter circuits can prevent tripping caused by the cumulative leakage from several tools or by leakages from extremely long circuits.

SCAFFOLDS

Scaffolding is defined as a temporary, elevated work platform used in construction. Many construction companies could not accomplish their tasks without the use of scaffolding. The following regulations and safe work practices will allow construction companies to use scaffolding safely.

General Requirements

- The footing or anchorage for scaffolds shall be sound, rigid, and capable of carrying the maximum intended load without settling or displacement. Unstable objects such as barrels, boxes, loose brick, or concrete blocks shall not be used to support scaffolds or planks.

- No scaffold shall be erected, moved, dismantled, or altered except under the supervision of a competent person.

- The fall-protection trigger height for scaffolding is 10 feet. Guardrails and toe boards shall be installed on all open sides and ends of platforms more than 10 feet above the ground or floor.

- Guardrails shall be 2 × 4 inches, or the equivalent, approximately 42 inches high, with midrails, when required. Supports shall be at intervals not to exceed 8 feet. Toe boards shall be a minimum of 4 inches in height.

- Where persons are required to work or pass under the scaffold, scaffolds shall be provided with a screen between the toe board and the guardrail, extending along the entire opening, consisting of No. 18 gauge U.S. Standard wire ½-inch mesh, or the equivalent.

- Scaffolds and their components shall be capable of supporting without failure at least four times the maximum intended load.

- Any scaffold including accessories such as braces, brackets, trusses, screw legs, ladders, etc., damaged or weakened from any cause shall be immediately repaired or replaced.

- All load-carrying timber members of scaffold framing shall be a minimum of 1,500 fiber (Stress Grade) construction-grade lumber. All dimensions are nominal sizes as provided in the American Lumber Standards, except that where rough sizes are noted, only rough or undressed lumber of the size specified will satisfy minimum requirements.

- All planking shall be Scaffold Grades, or equivalent, as recognized by approved grading rules for the species of wood used.

- The maximum permissible span for 1¼ × 9-inch or wider plank of full thickness shall be 4 feet with medium duty loading of 50 psf.

- All planking of platforms shall be overlapped (minimum 12 inches) or secured from movement.

- An access ladder or equivalent safe access shall be provided.

- Scaffold planks shall extend over their end supports not less than 6 inches nor more than 12 inches.

- The poles, legs, or uprights of scaffolds shall be plumb and securely and rigidly braced to prevent swaying and displacement.

- Overhead protection shall be provided for employees on a scaffold exposed to overhead hazards.

- Slippery conditions on scaffolds shall be eliminated as soon as possible after they occur.

- No welding, burning, riveting, or open flame work shall be performed on any staging suspended by means of fiber or synthetic rope. Only treated or protected fiber or synthetic ropes shall be used for or near any work involving the use of corrosive substances or chemicals.

- Wire, synthetic, or fiber rope used for scaffold suspension shall be capable of supporting at least six times the rated load.

- The use of shore or lean-to scaffolds is prohibited.

- Materials being hoisted onto a scaffold shall have a tag line.

- Employees shall not work on scaffolds during storms or high winds.

- Tools, materials, and debris shall not be allowed to accumulate in quantities to cause a hazard.

Tube-and-Coupler Scaffolds

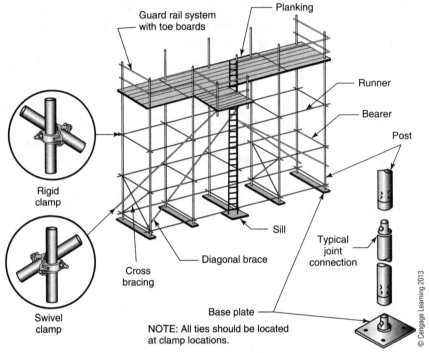

Tube-and-coupler scaffolding

- A light-duty tube-and-coupler scaffold shall have all posts, bearers, runners, and bracing of nominal 2-inch O.D. steel tubing. The posts shall be spaced no more than 6 feet apart by 10 feet along the length of the scaffold. Other structural metals when used must be designed to carry an equivalent load. No dissimilar metals shall be used together.

- A medium-duty tube-and-coupler scaffold shall have all posts, runners, and bracing of nominal 2 inch O.D. steel tubing. Posts spaced not more than 6 feet apart by 8 feet along the length of the scaffold shall have bearers of nominal 2½-inch O.D. steel tubing. Posts spaced not more than 5 feet apart by 8 feet along the length of the scaffold shall have bearers of nominal 2-inch O.D. steel tubing. Other structural metals, when used, must be designed to carry an equivalent load. No dissimilar metals shall be used together.

- A heavy-duty tube-and-coupler scaffold shall have all posts, runners, and bracing of nominal 2-inch O.D. steel tubing, with the posts spaced not more than 6 feet by 6 feet 6 inches. Other structural metals, when used, must be designed to carry an equivalent load. No dissimilar metals shall be used together.

- Tube-and-coupler scaffolds shall be limited in heights and working levels to those called out by OSHA or the manufacturer.

- All tube-and-coupler scaffolds shall be constructed and erected to support four times the maximum intended load.

- Posts shall be accurately spaced, erected on suitable bases, and maintained plumb.

- Runners shall be erected along the length of the scaffold, located on both the inside and the outside posts at even height. Runners shall be interlocked to the inside and the outside posts at even heights. Runners shall be interlocked to form continuous lengths and coupled to each post. The bottom runners shall be located as close to the base as possible. Runners shall be placed not more than 6 feet 6 inches on centers.

- Bearers shall be installed transversely between posts and shall be securely coupled to the posts bearing on the runner coupler. When coupled directly to the runners, the coupler must be kept as close to the posts as possible.

- Bearers shall be at least 4 inches but not more than 12 inches longer than the post spacing or runner spacing.

- Cross bracing shall be installed across the width of the scaffold at least every third set of posts horizontally and every fourth runner vertically. Such bracing shall extend diagonally from the inner and outer runners upward to the next outer and inner runners.

- Longitudinal diagonal bracing on the inner and outer rows of poles shall be installed at approximately a 45-deg. angle from near the base of the first outer post upward to the extreme top of the scaffold. Where the longitudinal length of the scaffold permits, such bracing shall be duplicated beginning at every fifth post. In a similar manner, longitudinal diagonal bracing shall also be installed from the last post extending back and upward toward the first post. Where conditions preclude the attachment of this bracing to the posts, it may be attached to the runners.

- The entire scaffold shall be tied to and securely braced against the building at intervals not to exceed 30 feet horizontally and 26 feet vertically.

- Guardrails, made of lumber not less than 2 × 4 inches (or other material providing equivalent protection), approximately 42 inches high, with midrails of 1 × 6-inch lumber (or other material providing equivalent protection), and toe boards shall be installed at all open sides and ends on all scaffolds more than 10 feet above the ground or floor. Toe boards shall be a minimum of 4 inches in height.

Tubular Welded Frame Scaffolds

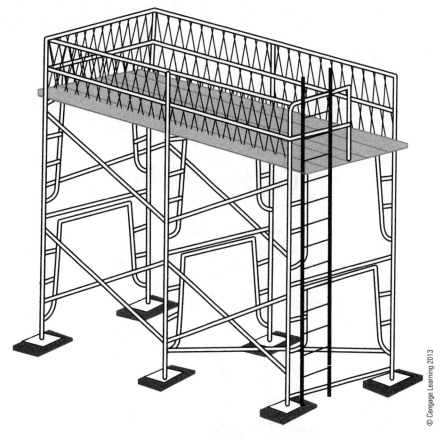

Tubular welded frame scaffolds

© Cengage Learning 2013

- Metal tubular frame scaffolds, including accessories such as braces, brackets, trusses, screw legs, ladders, etc., shall be designed, constructed, and erected to safely support four times the maximum rated load.
- Spacing of panels or frames shall be consistent with the loads imposed.

- Scaffolds shall be properly braced by cross bracing or diagonal braces, or both, for securing vertical members together laterally, and the cross braces shall be of such length as will automatically square and align vertical members so that the erected scaffold is always plumb, square, and rigid. All brace connections shall be made secure.

- Scaffold legs shall be set on adjustable bases or plain bases placed on mud sills or other foundations adequate to support the maximum rated load.

- The frames shall be placed one on top of the other with coupling or stacking pins to provide proper vertical alignment of the legs.

- Where uplift may occur, panels shall be locked together vertically by pins or other equivalent suitable means.

- To prevent movement, the scaffold shall be secured to the building or structure at intervals not to exceed 30 feet horizontally and 26 feet vertically.

- Maximum permissible spans or planking shall be in conformity with OSHA regulations.

- Drawings and specifications for all frame scaffolds over 125 feet in height above the base plates shall be designed by a registered professional engineer.

- Guardrails made of lumber, not less than 2 × 4 inches (or other material providing equivalent protection), and approximately 42 inches high, with midrails of 1 × 6-inch lumber (or other material providing equivalent protection), and toe boards shall be installed at all open sides and ends on all scaffolds more than 10 feet above the ground or floor. Toe boards shall be a minimum of 4 inches in height.

Manually Propelled Mobile Scaffolds

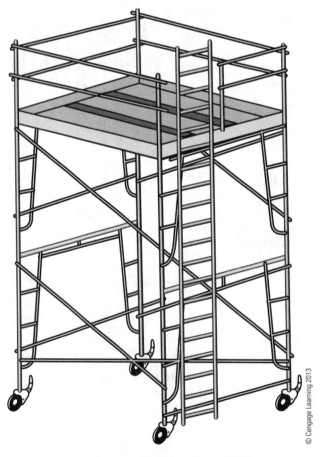

© Cengage Learning 2013

Manually propelled mobile scaffolds

- When free-standing mobile scaffold towers are used, the height shall not exceed four times the minimum base dimension.
- Casters shall be properly designed for strength and dimensions to support four times the maximum intended load. All casters shall be provided with a positive locking device to hold the scaffold in position.
- Scaffolds shall be properly braced by cross bracing and horizontal bracing.
- Platforms shall be tightly planked for the full width of the scaffold except for necessary entrance openings. Platforms shall be secured in place.
- A ladder on stairway shall be provided for proper access and exit and shall be affixed or built into the scaffold and so located that when in use it will not have a tendency to tip the scaffold. A landing platform must be provided at intervals not to exceed 35 feet.
- The force necessary to move the mobile scaffold shall be applied near or as close to the base as practicable and provision shall be made to stabilize the tower during movement from one location to another. Scaffolds shall only be moved on level floors, free of obstructions and openings.
- The employer shall not allow employees to ride on manually propelled scaffolds unless the following conditions exist:
  - The floor or surface is within 3 degrees of level and free from pits, holes, or obstructions.
  - The minimum dimension of the scaffold base when ready for rolling is at least one-half of the height. Outriggers, if used, shall be installed on both sides of the staging.
  - The wheels are equipped with rubber or similar resilient tires.
  - All tools and materials are secured or removed from the platform before the mobile scaffold is moved.
- Scaffolds in use by any persons shall rest upon a suitable footing and shall stand plumb. The casters or wheels shall be locked to prevent any movement.
- Guardrails made of lumber, not less than 2 × 4 inches (or other material providing equivalent protection), approximately 42 inches high, with midrails, of 1 × 6-inch lumber (or other material providing equivalent protection), and toe boards shall be installed at all open sides and ends on all scaffolds more than 10 feet above the ground or floor. Toe boards shall be a minimum of 4 inches in height.

Masons' Adjustable Multiple-Point Suspension Scaffolds

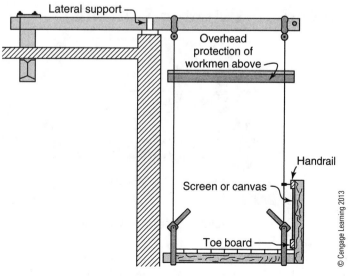

Mason's adjustable multiple-point suspension scaffold

- The scaffold shall be capable of sustaining a working load of 50 pounds per square foot and shall not be loaded in excess of that figure.
- The scaffold shall be provided with hoisting machines that meet the requirements of Underwriters' Laboratories or Factory Mutual Engineering Corporation.
- The platform shall be supported by wire ropes capable of supporting at least six times the intended load, suspended from overhead outrigger beams.
- The scaffold outrigger beams shall consist of structural metal securely fastened or anchored to the frame or floor system of the building or structure.
- Each outrigger beam shall be equivalent in strength to at least a standard 7-inch, 15.3-pound steel I-beam, at least 15 feet long, and shall not project more than 6 feet 6 inches beyond the bearing point.
- Where the overhang exceeds 6 feet 6 inches, outrigger beams shall be composed of stronger beams or multiple beams and be installed under the supervision of a competent person.
- All outrigger beams shall be set and maintained with their webs in a vertical position.
- A stop bolt shall be placed at each end of every outrigger beam.
- The outrigger beam shall rest on suitable wood bearing blocks.
- The free end of the suspension wire ropes shall be equipped with proper size thimbles and secured by splicing or other equivalent means. The running ends shall be securely attached to the hoisting drum and at least four turns of wire rope shall at all times remain on the drum. The use of fiber rope is prohibited.
- Where a single outrigger beam is used, the steel shackles or clevises with which the wire ropes are attached to the outrigger beams shall be placed directly over the hoisting drums.
- The scaffold platform shall be equivalent in strength to at least 2-inch planking.
- When employees are at work on the scaffold and an overhead hazard exists, overhead protection shall be provided on the scaffold, not more than 9 feet above the platform, consisting of 2-inch planking, or material of equivalent strength, laid tight and extending not less than the width of the scaffold.
- Each scaffold shall be installed or relocated under the supervision of a competent person.
- Guardrails made of lumber, not less than 2 × 4 inches (or other material providing equivalent protection), approximately 42 inches high, with midrails, and toe boards shall be installed at all open sides and ends on all scaffolds more than 10 feet above the ground or floor. Toe boards shall be a minimum of 4 inches in height.

Two-Point Suspension (Swinging Scaffolds)

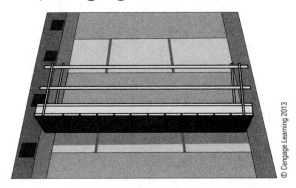

© Cengage Learning 2013

Two-point suspension/swinging scaffold

- Two-point suspension scaffold platforms shall be not less than 20 inches or more than 36 inches wide overall. The platform shall be securely fastened to the hangers by U-bolts or by other equivalent means.

- The hangers of two-point suspension scaffolds shall be made of mild steel, or other equivalent materials, having a cross-sectional area capable of sustaining four times the maximum rated load and shall be designed with a support for guardrail, intermediate rail, and toe board.

- When hoisting machines are used on two-point suspension scaffolds, such machines shall be of a design tested and approved by Underwriters' Laboratories or Factory Mutual Engineering Corporation.

- The roof irons or hooks shall be of mild steel, or other equivalent material, of proper size and design, securely installed and anchored. Tiebacks of ¾-inch manila rope, or the equivalent, shall serve as a secondary means of anchorage, installed at right angles to the face of the building, whenever possible, and secured to a structurally sound portion of the building.

- Two-point suspension scaffolds shall be suspended by wire, synthetic, or fiber ropes capable of supporting at least six times the rated load. All other components shall be capable of supporting at least four times the rated load.

- The sheaves of all blocks, consisting of at least one double and one single block, shall fit the size and type of rope used.

- All wire ropes, fiber and synthetic ropes, slings, hangers, platforms, and other supporting parts shall be inspected before every installation. Periodic inspections shall be made while the scaffold is in use.

- On suspension scaffolds designed for a working load of 500 pounds, no more than two employees shall be permitted to work at one time. On suspension scaffolds with a working load of 750 pounds, no more than three employees shall be permitted to work at one time. Each employee shall be protected by an approved safety life belt attached to a lifeline. The lifeline shall be securely attached to substantial members of the structure (not scaffold), or to securely rigged lines, which will safely suspend the employee in case of a fall. In order to keep the lifeline continuously attached, with a minimum of slack, to a fixed structure, the attachment point of the lifeline shall be appropriately changed as the work progresses.

- Two-point suspension scaffolds shall be securely lashed to the building or structure to prevent them from swaying. Window cleaners' anchors shall not be used for this purpose.

- The platform of every two-point suspension scaffold shall be one of the following types:
  - **Ladder-type platforms.** The side stringer shall be of clear straight-grained spruce or materials of equivalent strength and durability. The rungs shall be of straight-grained oak, ash, or hickory, at least 1⅛ inch in diameter, with ⅞-inch tenons mortised into the side stringers at least ⅞ inch. The stringers shall be tied together with tie rods not less than ¼ inch in diameter, passing through the stringers and riveted up tight against washers on both ends. The flooring strips shall be spaced not more than ⅝ inch apart except at the side rails where the space may be 1 inch.
  - **Plank-type platforms.** Plank-type platforms shall be composed of not less than nominal 2 × 10-inch unspliced planks, properly cleated together on the underside, starting 6 inches from each end; intervals in between shall not exceed 4 feet. The plank-type platform shall not extend beyond the hangers more than 12 inches. A bar or other effective means shall be securely fastened to the platform at each end to prevent its slipping off the hanger. The span between hangers for plank-type platforms shall not exceed 8 feet.
  - **Beam-type platforms.** Beam platforms shall have side stringers of lumber not less than 2 × 6 inches set on edge. The span between hangers shall not exceed 12 feet when beam platforms are

used. The flooring shall be supported on 2 × 6-inch cross beams, laid flat and set into the upper edge of the stringers with a snug fit, at intervals of not more than 4 feet, securely nailed in place. The flooring shall be of 1 × 6-inch material properly nailed. Floor boards shall not be spaced more than ½ inch apart.

- ○ **Light metal-type platforms,** when used, shall be tested and listed according to Underwriters' Laboratories or Factory Mutual Engineering Corporation.
- • Guardrails made of lumber, not less than 2 × 4 inches (or other material providing equivalent protection), approximately 42 inches high, with a midrail, and toe boards shall be installed at all open sides and ends on all scaffolds more than 10 feet above the ground or floor. Toe boards shall be a minimum of 4 inches in height.

Plasterers', Decorators', and Large-Area Scaffolds

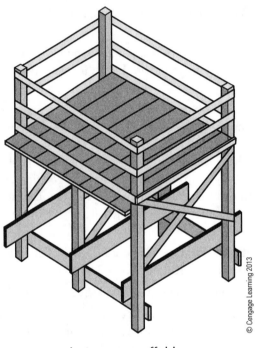

© Cengage Learning 2013

Large-area scaffold

- • Plasterers', lathers', and ceiling workers' inside scaffolds shall be constructed in accordance with the general requirements set forth for independent wood pole scaffolds.
- • All platform planks shall be laid with the edges close together.
- • When independent pole scaffold platforms are erected in sections, such sections shall be provided with connecting runways equipped with substantial guardrails.
- • Guardrails made of lumber, not less than 2 × 4 inches (or other material providing equivalent protection), approximately 42 inches high, with midrails of 1 × 6-inch lumber (or other material providing equivalent protection), and toe boards shall be installed on all open sides and ends of all scaffolds more than 10 feet above the ground or floor. Toe boards shall be a minimum of 4 inches in height.

Ladder Jack Scaffolds

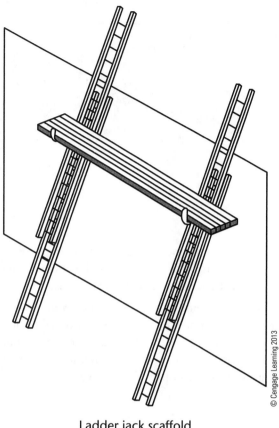

Ladder jack scaffold

- All ladder jack scaffolds shall be limited to light duty and shall not exceed a height of 20 feet above the floor or ground.
- All ladders used in connection with ladder jack scaffolds shall be heavy-duty ladders and shall be designed and constructed in accordance with American National Standards Institute A 14.1-1968, Safety Code for Portable Wood Ladders, and A 14.2-1968, Safety Code for Portable Metal Ladders. Cleated ladders shall not be used for this purpose.
- The ladder jack shall be so designed and constructed that it will bear on the side rails in addition to the ladder rungs, or if bearing on rungs only, the bearing area shall be at least 10 inches on each rung.
- Ladders used in conjunction with ladder jacks shall be so placed, fastened, held, or equipped with devices as to prevent slipping.
- The wood platform planks shall be not less than 2 inches nominal in thickness. Both metal and wood platform planks shall overlap the bearing surface not less than 12 inches. The span between supports for wood shall not exceed 8 feet. Platform width shall be not less than 18 inches.
- Not more than two employees shall occupy any given 8 feet of any ladder jack scaffold at any one time.

Form Scaffolds

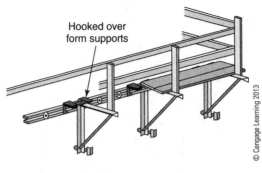

Form scaffold

- Form scaffolds shall be constructed of wood or other suitable materials, such as steel or aluminum members of known strength characteristics. All scaffolds shall be designed and erected with a minimum safety factor of four, computed on the basis of the maximum rated load.

- All scaffold planking shall be a minimum of 2 × 10-inch nominal Scaffold Grade, as recognized by approved grading rules for the species of lumber used, or equivalent material. Maximum permissible spans shall not exceed 8 feet on centers for 2 × 10-inch nominal planking. Scaffold planks shall be either nailed or bolted to the ledgers or of such length that they overlap the ledgers at least 6 inches. Unsupported projecting ends of scaffolding planks shall be limited to a maximum overhang of 12 inches.

- Scaffolds shall not be loaded in excess of the working load for which they were designed.

- Figure-four form scaffolds:
  - Figure-four scaffolds are intended for light duty and shall not be used to support loads exceeding 25 pounds per square foot unless specifically designed for heavier loading.
  - Figure-four form scaffold frames shall be spaced not more than 8 feet on centers and constructed from sound lumber, as follows: The outrigger ledger shall consist of two pieces of 1 × 6-inch or heavier material nailed on opposite sides of the vertical form support. Ledgers shall project not more than 3 feet 6 inches from the outside of the form support and shall be substantially braced and secured to prevent tipping or turning. The knee or angle brace shall intersect the ledger at least 3 feet from the form at an angle of approximately 45 deg., and the lower end shall be nailed to a vertical support. The platform shall consist of two or more 2 × 10-inch planks, which shall be of such length that they extend at least 6 inches beyond ledgers at each end unless secured to the ledgers. When planks are secured to the ledgers (nailed or bolted), a wood filler strip shall be used between the ledgers. Unsupported projecting ends of planks shall be limited to an overhang of 12 inches.

- Metal bracket form scaffolds:
  - Metal brackets or scaffold jacks that are an integral part of the form shall be securely bolted or welded to the form. Folding-type brackets shall be either bolted or secured with a locking-type pin when extended for use.
  - "Clip-on" or "hook-over" brackets may be used, provided the form walers are bolted to the form or secured by snap ties or shea-bolt extending through the form and securely anchored.
  - Metal brackets shall be spaced not more than 8 feet on centers.

Scaffold planks shall be either bolted to the metal brackets or of such length that they overlap the brackets at each end by at least 6 inches. Unsupported projecting ends of scaffold planks shall be limited to a maximum overhang of 12 inches.

Metal bracket form scaffolds shall be equipped with wood guardrails, intermediate rails, toe boards, and scaffold planks meeting the minimum dimensions.

- Wooden bracket form scaffolds:

  ○ Wooden bracket form scaffolds shall be an integral part of the form panel. Scaffold planks shall be either nailed or bolted to the ledgers or of such length that they overlap the ledgers at each end by at least 6 inches. Unsupported projecting ends of scaffold planks shall be limited to a maximum overhang of 12 inches.

  ○ Guardrails and toe boards shall be installed on all open sides and ends of platforms and scaffolding over 10 feet above the floor or ground. Guardrails shall be made of lumber 2 × 4-inch nominal dimension (or other material providing equivalent protection), approximately 42 inches high, supported at intervals not to exceed 8 feet. Guardrails shall be equipped with midrails constructed of 1 × 6-inch nominal lumber (or other material providing equivalent protection). Toe boards shall extend not less than 4 inches above the scaffold plank.

Pump Jack Scaffolds

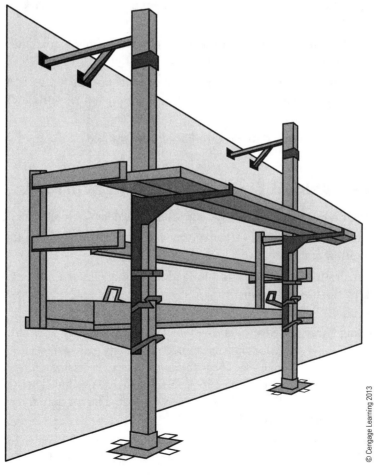

© Cengage Learning 2013

Pump jack scaffold

- Pump jack scaffolds shall not carry a working load exceeding 500 pounds; and be capable of supporting without failure at least four times the maximum intended load. The manufactured components shall not be loaded in excess of the manufacturer's recommended limits.

- Pump jack brackets, braces, and accessories shall be fabricated from metal plates and angles. Each pump jack bracket shall have two positive gripping mechanisms to prevent any failure or slippage.

- The platform bracket shall be fully decked and the planking secured.

- When wood scaffold planks are used as platforms, poles used for pump jacks shall not be spaced more than 10 feet center to center.

- Poles shall not exceed 30 feet in height, unless manufactured poles allow for greater heights.

- Poles shall be secured to the work wall by rigid triangular bracing, or equivalent, at the bottom, top, and other points as necessary, to provide a maximum vertical spacing of not more than 10 feet between braces. Each brace shall be capable of supporting a minimum of 225 pounds tension or compression.

- For the pump jack bracket to pass bracing already installed, an extra brace shall be used approximately 4 feet above the one to be passed until the original brace is reinstalled.

- All poles shall bear on mud sills or other adequate firm foundations.

- A ladder, in accordance with OSHA regulations, shall be provided for access to the platform during use.

- Not more than two persons shall be permitted at one time upon a pump jack scaffold between any two supports.

- Pump jack scaffolds shall be provided with standard guardrails, but no guardrail is required when safety harnesses with lifelines are provided for employees.

- When a workbench is used at an approximate height of 42 inches, the top guardrail may be eliminated, if the workbench is fully decked, the planking secured, and is capable of withstanding 200 pounds force in any direction.

- Employees shall not be permitted to use a workbench as a scaffold platform.

Mobile Tubular Welded Sectional Folding Scaffolds

- **General:** Units including sectional stairway and sectional ladder scaffolds shall be designed properly.

- **Stairway:** An integral stairway and work platform shall be incorporated into the structure of each sectional folding stairway scaffold.

- **Bracing:** An integral set of pivoting and hinged folding diagonal and horizontal braces and a detachable work platform shall be incorporated into the structure of each sectional folding ladder scaffold.

- **Sectional Folding Stairway Scaffolds:** Sectional folding stairway scaffolds shall be designed as medium-duty scaffolds except for high clearance. These special base sections shall be designed as light-duty scaffolds. When upper sectional folding stairway scaffolds are used with a special high-clearance base, the load capacity of the entire scaffold shall be reduced accordingly. The width of a sectional folding stairway scaffold shall not exceed 4½ feet. The maximum length of a sectional folding stairway scaffold shall not exceed 6 feet.

- **Sectional Folding Ladder Scaffolds:** Sectional folding ladder scaffolds shall be designed as light-duty scaffolds including special base (open end) sections that are designed for high clearance. For certain

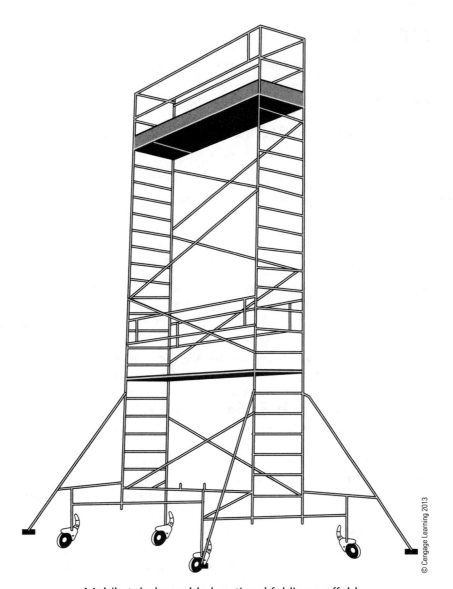

Mobile tubular welded sectional folding scaffold

special applications, the 6-foot folding ladder scaffolds, except for special high-clearance base sections, shall be designed for use as medium-duty scaffolds. The width of a sectional folding ladder scaffold shall not exceed 4½ feet. The maximum length of a sectional folding ladder scaffold shall not exceed 6 feet 6 inches for a 6-foot-long unit, 8 feet 6 inches for an 8-foot unit, or 10 feet 6 inches for a 10-foot-long unit.

- **End Frames:** The end frames of sectional ladder and stairway scaffolds shall be designed so that the horizontal bearers provide supports for multiple planking levels.

- **Erection:** Only the manufacturer of the scaffold or his qualified designated agent shall be permitted to erect or supervise the erection of scaffolds exceeding 50 feet in height above the base, unless such structure is approved in writing by a licensed professional engineer, or erected in accordance with instructions furnished by the manufacturer.

FALL PROTECTION

In the U.S. construction industry, falls are the leading cause of worker fatalities. Each year, on average, between 150 and 200 workers are killed and more than 100,000 are injured as a result of falls at construction sites. OSHA recognizes that accidents involving falls are generally complex events frequently involving a variety of factors. Consequently, the standard for fall protection deals with both the human and equipment-related issues in protecting workers from fall hazards. For example, employers and employees need to do the following:

- Where protection is required, select fall protection systems appropriate for given situations.
- Use proper construction and installation of safety systems.
- Supervise employees properly.
- Use safe work practices.
- Train workers in the proper selection, use, and maintenance of all protection systems.

This regulation covers most construction workers except those inspecting, investigating, or assessing workplace conditions prior to the actual start of work or after all work has been completed.

Fall protection is needed in multiple situations. These include, but are not limited to, ramps, runways, and other walkways; excavations; hoist areas; holes; formwork and reinforcing steel; leading edge work; unprotected sides and edges; overhand bricklaying and related work; roofing work; precast concrete erection; wall openings; residential construction; and other walking/working surfaces. The rule sets a uniform threshold height of 6 feet, thereby providing consistent protection. This means that construction employers must protect their employees from fall hazards and falling objects whenever an affected employee is 6 feet or more above a lower level. Protection must also be provided for construction workers who are exposed to the hazard of falling into dangerous equipment.

Under this regulation, employers are able to select fall protection measures compatible with the type of work being performed. Fall protection generally can be provided through the use of guardrail systems, safety net systems, personal fall arrest systems, positioning-device systems, and warning-line systems, among others.

The OSHA rule clarifies what an employer must do to provide fall protection for employees, such as identifying and evaluating fall hazards and providing specific training. Requirements to provide fall protection for workers on scaffolds and ladders and for workers engaged in building steel erection are covered in other subparts of OSHA regulations.

The standard prescribes the duty to provide fall protection, sets the criteria and practices for fall protection systems, and requires training. It covers hazard assessment and fall protection and safety monitoring systems. Also addressed are controlled access zones, safety nets, and guardrail, personal fall arrest, warning-line, and positioning-device systems.

Duty to Have Fall Protection

Employers are required to assess the workplace to determine if the walking/working surfaces on which employees are to work have the strength and structural integrity to safely support workers. Employees are not permitted to work on those surfaces until it has been determined that the surfaces have the requisite strength and structural integrity to support the workers. Once employers have determined that the surface is safe for employees to work on, the employer must select one of the options listed for the work operation if a fall hazard is present.

For example, if an employee is exposed to falling 6 feet or more from an unprotected side or edge, the employer must select either a guardrail system, safety net system, or personal fall arrest system to protect the worker. Similar requirements are prescribed for other fall hazards as follows.

Controlled Access Zone

A controlled access zone is a work area designated and clearly marked in which certain types of work (such as overhand bricklaying) may take place without the use of conventional fall protection systems—guardrail, personal-arrest, or safety net—to protect the employees working in the zone.

Controlled access zones are used to keep out workers other than those authorized to enter work areas from which guardrails have been removed. Where there are no guardrails, masons are the only workers allowed in controlled access zones.

Controlled access zones, when created to limit entrance to areas where leading edge work and other operations are taking place, must be defined by a control line or by any other means that restrict access. Control lines shall consist of ropes, wires, tapes or equivalent materials, and supporting stanchions, and each must be:

- Flagged or otherwise clearly marked at not more than 6-foot intervals with high-visibility material
- Rigged and supported in such a way that the lowest point (including sag) is not less than 39 inches from the walking/working surface and the highest point is not more than 45 inches nor more than 50 inches when overhand bricklaying operations are being performed—from the walking/working surface
- Strong enough to sustain stress of not less than 200 pounds. Control lines shall extend along the entire length of the unprotected or leading edge and shall be approximately parallel to the unprotected or leading edge
- Control lines also must be connected on each side to a guardrail system or wall

When control lines are used, they shall be erected not less than 6 feet nor more than 25 feet from the unprotected or leading edge, except when precast concrete members are being erected. In the latter case, the control line is to be erected not less than 6 feet nor more than 60 feet or half the length of the members being erected, whichever is less, from the leading edge.

Controlled access zones when used to determine access to areas where overhand bricklaying and related work are taking place are to be defined by a control line erected not less than 10 feet nor more than 15 feet from the working edge. Additional control lines must be erected at each end to enclose the controlled access zone. Only employees engaged in overhand bricklaying or related work are permitted in the controlled access zones.

On floors and roofs where guardrail systems are not in place prior to the beginning of overhand bricklaying operations, controlled access zones will be enlarged as necessary to enclose all points of access, material handling areas, and storage areas. On floors and roofs where guardrail systems are in place, but need to be removed to allow overhand bricklaying work or leading edge work to take place, only that portion of the guardrail necessary to accomplish that day's work shall be removed.

Fall Protection at Excavations

Each employee at the edge of an excavation 6 feet or more deep shall be protected from falling by guardrail systems, fences, barricades, or covers when the excavations are not readily seen because of plant growth or other visual barrier. Where walkways are provided to permit employees to cross over excavations, guardrails are required on the walkway if it is 6 feet or more above the excavation.

Formwork and Reinforcing Steel

For employees, while moving vertically and/or horizontally on the vertical face of rebar assemblies built in place, fall protection is not required when employees are moving. OSHA considers the multiple hand holds and foot holds on rebar assemblies as providing similar protection as that provided by a fixed ladder; consequently, no fall protection is necessary while moving point to point for heights below 24 feet. An employee must be provided with fall protection when climbing or otherwise moving at a height more than 24 feet, the same as for fixed ladders.

Hoist Areas

Each employee in a hoist area shall be protected from falling 6 feet or more by guardrail systems or personal fall arrest systems. If guardrail systems (or chain gate or guardrail) or portions thereof must be removed to facilitate hoisting operations, as during the landing of materials, and a worker must lean through the access opening or out over the edge of the access opening to receive or guide equipment and materials, that employee must be protected by a personal fall arrest system.

Holes

Personal fall arrest systems, covers, or guardrail systems shall be erected around holes (including skylights) that are more than six feet above lower levels.

Leading Edges

Each employee performing overhand bricklaying and related work six feet or more above lower levels shall be protected by guardrail systems, safety net systems, or personal fall arrest systems, or shall work in a controlled access zone. All employees reaching more than 10 inches below the level of a walking/working surface on which they are working shall be protected by a guardrail system, safety net system, or personal fall arrest system.

Precast Concrete Erection

Each employee who is six feet or more above lower levels while erecting precast concrete members and related operations such as grouting of precast concrete members shall be protected by guardrail systems, safety net systems, or personal fall arrest systems. Where the employer can demonstrate, however, that it is infeasible or creates a greater hazard to use those systems, the employer must develop and implement a fall protection plan that meets the requirements of 29 CFR 1926.502(k). An example of a fall protection plan for Precast Concrete Erection can be found in Resource J located at the back of this book and at dewalt.com/guides. IMPORTANT: OSHA wants companies to use conventional fall protection at all times. The use of a fall protection plan is only recommended where the employers can **prove** that the use of conventional fall protection is infeasible or creates a greater hazard.

Residential Construction

OSHA has changed the rules for fall protection within the residential construction industry. As of June 16, 2011, the alternative fall protection procedures for residential construction are no longer allowed, unless the company can prove that conventional fall protection systems (guardrail systems, personal fall arrest systems, safety net systems, and covers) are infeasible or create a greater hazard. OSHA believes that most practices during residential construction can be performed with conventional fall protection.

Specific Fall Protection Practices for Residential Construction

The following are some options available for fall protection practices for use during residential construction:

Installing Roof Trusses

Numerous methods can be used to prevent fall-related injuries and fatalities among workers installing roof trusses.

Bracket Scaffold   A bracket scaffold can be placed on the interior or exterior of a structure. The scaffold can provide a stable working platform. When bracket scaffolds are used on the interior of the structure, the exterior wall can limit employee exposures to fall hazards.

Bracket scaffolds

Exterior bracket scaffolds can also be used for installing roof trusses and other rooftop construction activities. The guardrail system on the scaffold can provide fall protection. With the addition of toeboards, falling object protection can be provided to the areas below.

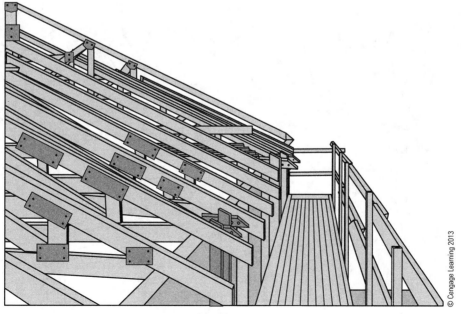

Exterior brackets

Ladders   Platform ladders and step ladders can provide a stable, elevated platform from which to work.

Step ladder

Anchors   A spreader braces the trusses and distributes arrest forces across several trusses. Spreaders can act as anchors for personal fall arrest systems and fall restraint. They can be reused according to the manufacturer's instructions. It is important to refer to the truss manufacturer's instructions and have a qualified person determine if trusses will meet strength requirements for a personal fall arrest system or fall restraint system.

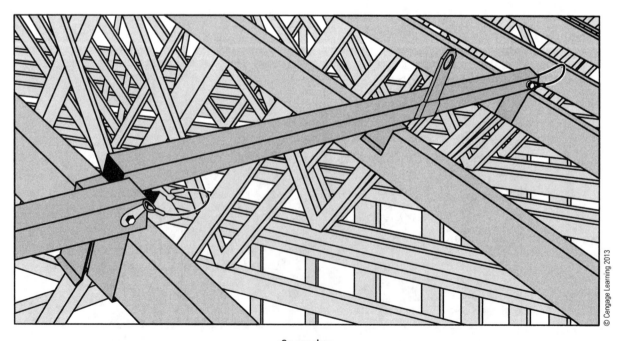

Spreader

Installing Ridge Poles and Rafters

Although the use of roof trusses is nearly universal, some builders still frame roof systems with ridge poles and rafters. While performing this task, workers need to be protected from falls.

Anchors   Employees installing ridge poles and rafters can use strap anchors and bolt-on anchors. These anchors can be used with personal fall arrest systems and fall restraint to provide fall protection for workers engaged in this activity. Both anchors can be removed and reused according to the manufacturer's instructions.

Anchors

Installing Roof Sheathing

Once a roof has been framed, roof sheathing operations can begin. There are serious fall hazards associated with this activity, but there are a number of ways to protect workers.

Safety Net System   Safety net systems can be used as fall protection for workers installing roof sheathing.

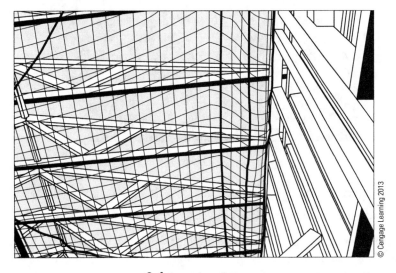

Safety net system

Bracket Scaffold   A bracket scaffold can be attached to the top plate of a structure. The scaffold can provide a secure work platform from which to install roof sheathing.

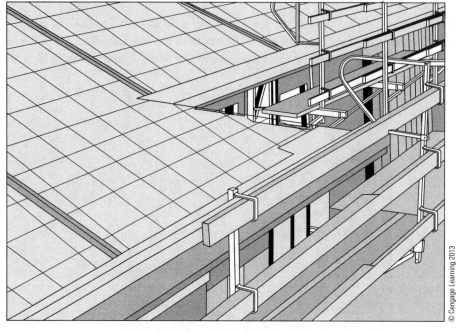

Bracket scaffold

Anchors   Anchors and retractable lifeline stands can be used by workers installing roof sheathing.

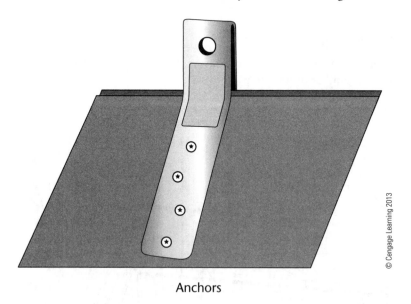

Anchors

Roofing—Weatherproofing

As with other roofing activities, fall protection is critical for this type of work.

Bracket Scaffolds   An exterior bracket scaffold can be used for workers weatherproofing a roof. Bracket scaffolds can be especially useful for installing materials along the edge of the roof. Some exterior bracket scaffolds can be used as catch platforms to prevent workers from falling 6 feet to the lower level.

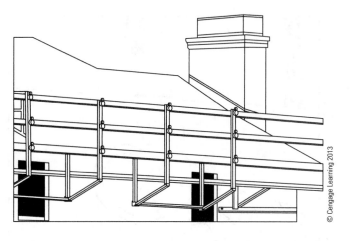

Bracket scaffold—roofing

Anchors   Permanent anchors can be installed during roofing operations and left in place after construction is complete. They can provide an anchorage point during the life of the roof. Reusable anchors can also be used while weatherproofing a roof. It is important to inspect these anchors prior to use.

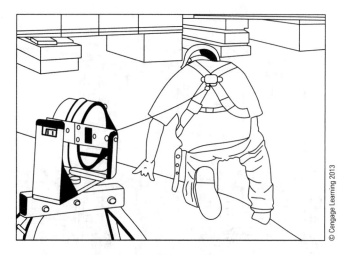

Permanent anchors

Foundation Walls and Formwork

In most residential construction, concrete or masonry block is used to create the foundation and the foundation walls of a structure. The concrete is usually poured into an excavation to create the foundation and the foundation walls.

Anchors   Anchors can be added to cured concrete. Anchors with expandable bolts can be placed in holes that have been drilled into the concrete.

Strap anchors can provide versatility and options for anchorage points while performing this type of work. Strap anchors can be looped over rebar and removed when no longer necessary.

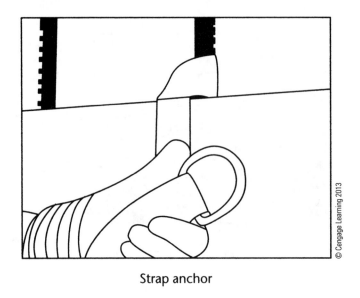

Strap anchor

Scaffolds   Scaffolds can provide elevated work surfaces for workers performing foundation work.

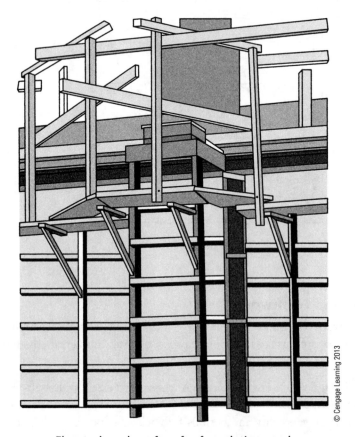

Elevated work surface for foundation work

Installing Floor Joists and Floor Trusses

Floor joists and floor trusses are usually constructed directly over the foundation walls. Fall hazards may be present, for example, if the structure being built has a basement.

Anchors   A reusable floor truss anchor can act as a temporary truss brace and spacer as well as an anchor point for a self-retracting lifeline. The device can spread shock loads over multiple trusses. These kinds of anchors can be uninstalled, moved, reinstalled, and reused as per the manufacturer's instructions.

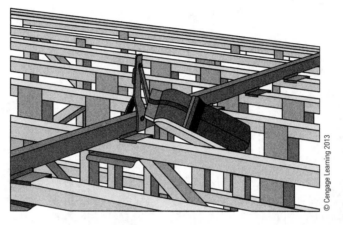

Reusable floor truss anchor

Scaffolds   Scaffolds can be used for residential construction workers installing floor joists and floor trusses.

Mobile scaffolds can be used effectively for residential construction workers. These scaffolds can be placed on the cured concrete basement floor of a structure. From the elevated platforms of the mobile scaffold, workers can install carrier beams, floor joists, and floor trusses.

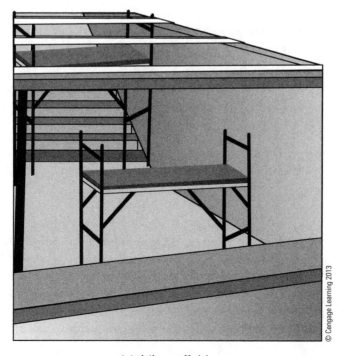

Mobile scaffolds

Wall bracket scaffolds can be used on a residential structure once a wall has been completed. These scaffolds can provide access around the perimeter of the structure and can be used by workers while they install carrier beams, floor joists, and floor trusses. This type of scaffold can also be used in other phases of residential construction.

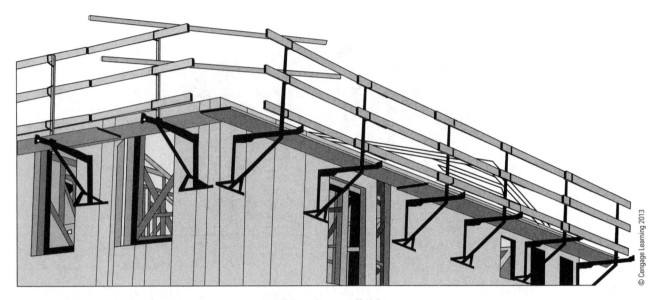

Wall bracket scaffolds

Installing Subfloors

Subfloors are usually installed by fastening a wood deck to floor joists and floor trusses. Because of the openings between floor joists and floor trusses, fall hazards may exist while performing this task.

Anchors   A truss bracket anchorage system can distribute the arresting forces across multiple trusses in the event of a fall. When appropriately installed in accordance with the manufacturer's instructions, these anchors can be used with personal fall arrest systems and fall restraints. Because these anchors are reusable, they can be uninstalled and reinstalled in accordance with the manufacturer's instructions.

Guardrails   Guardrail systems can be used to protect workers from falls during the performance of flooring and subflooring activities. If installed with a toeboard, guardrail systems can also protect workers on lower levels from falling objects.

Installing Walls

Framed wall sections usually are constructed on the ground and typically include sheathing and openings for windows and doors. Guardrails across these openings can help prevent falls while work is being performed in the house after the walls have been erected. These walls can be erected by using a lifting device such as a crane, boom truck, or forklift. Jacks can also be used to raise these walls. These practices greatly reduce the likelihood that a worker will be exposed to a fall during this stage of construction. For workers exposed to falls while framing walls, there remain various ways of protecting against falls to lower levels.

Truss bracket anchor

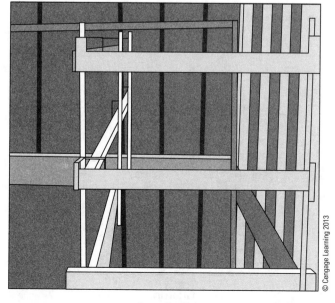

Guardrails

Anchors   Some models of strap anchors are looped through soft eyes or larger D rings and can be uninstalled, moved, reinstalled, and reused following the manufacturer's instructions.

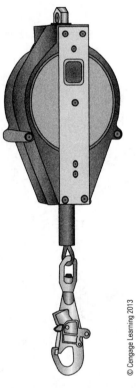

Self retracting lifeline

Strap anchors can also be preinstalled to provide an anchorage point once the walls have been erected and braced. They can be removed by cutting the strap off or pulling out the nails. Once cut or nailed, these must not be reused without being refurbished by the manufacturer.

Guardrails   Guardrail systems can provide fall protection while workers are framing walls.

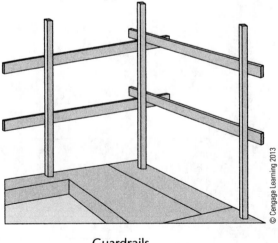

Guardrails

Sheathing Walls

Although it is common for sheathing to be included on panelized walls used in residential construction, sheathing still takes place on residential construction sites. Erecting the walls by lifting devices or jacks can lessen a worker's exposure to fall hazards.

Scaffolds   Welded End Frame (X brace) scaffolds (like tube/coupler and systems scaffolds) stand on their own. With proper bracing, the frames can be stacked from cellar hole to chimney top. Like other scaffolds, these can make stable work platforms for many residential construction operations. These stand-alone scaffolds can be used for sheathing and various other vertical wall operations such as masonry wall construction, siding, and stucco application.

Exterior bracket scaffolds can provide a work surface from which to attach sheathing to the frame.

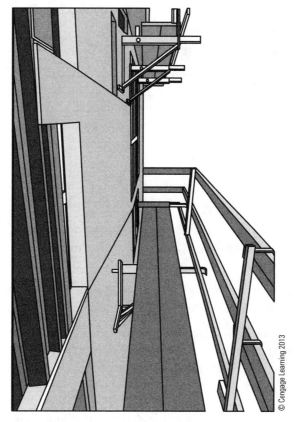

© Cengage Learning 2013

Exterior bracket scaffolds

A ladder jack scaffold consists of a platform resting on brackets attached to a ladder. Ladder jacks provide limited elevation and are primarily used in light applications because of their portability and cost-effectiveness.

Aerial Lifts   Aerial lifts may be used for wall sheathing work. In particular, boom-type elevating and rotating aerial work platforms can offer positioning flexibility and provide stable elevated platforms.

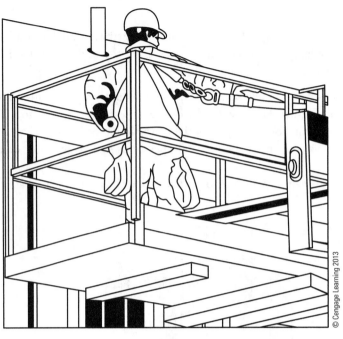

Aerial lift

© Cengage Learning 2013

Exterior Finishing

The exterior finishing phase of residential construction includes a number of activities, such as installing windows, doors, siding, and gutters. Many of these tasks pose fall hazards.

Aerial Lifts   Aerial lifts can be ideal equipment for exterior finishing. By providing a stable, level work surface and positioning flexibility, an aerial lift can be used for numerous activities associated with finishing the exterior of a residential structure.

Ladders   Ladders can provide access to areas of a structure where exterior finishing work will occur. They can provide a stable work surface for various exterior finishing tasks.

Scaffolds   Pump jack scaffolds consist of a platform supported by moveable brackets on vertical poles. Pump jacks are appealing for certain applications because they are easily adjusted to variable heights and are relatively inexpensive. They can include a material shelf for carrying supplies and tools.

Interior Finishing

Although much of this work may take place while exterior finishing is happening, interior finishing is one of the last phases of residential construction. This does not mean that fall hazards are no longer present.

Guardrails   Guardrails can be an excellent option for providing fall protection for work on or near stairways and landings. The addition of a toeboard can also prevent objects from falling to lower levels.

Ramps, Runways, and Other Walkways

Each employee using ramps, runways, and other walkways shall be protected from falling 6 feet or more by guardrail systems.

Roofing

Low-slope Roofs

Each employee engaged in roofing activities on low-slope roofs with unprotected sides and edges 6 feet or more above lower levels shall be protected from falling by guardrail systems, safety net systems, personal fall arrest systems, or a combination of a warning-line system and guardrail system, warning-line system and safety net system, warning-line system and personal fall arrest system, or warning-line system and safety monitoring system. On roofs 50 feet or less in width, the use of a safety monitoring system without a warning-line system is permitted.

Steep Roofs

Each employee on a steep roof with unprotected sides and edges 6 feet or more above lower levels shall be protected by guardrail systems with toe boards, safety net systems, or personal fall arrest systems.

Wall Openings

Each employee working on, at, above, or near wall openings (including those with chutes attached) where the outside bottom edge of the wall opening is 6 feet or more above lower levels and the inside bottom edge of the wall opening is less than 39 inches above the walking/working surface must be protected from falling by the use of a guardrail system, a safety net system, or a personal fall arrest system.

Fall Protection Systems Criteria and Practices

Guardrail Systems

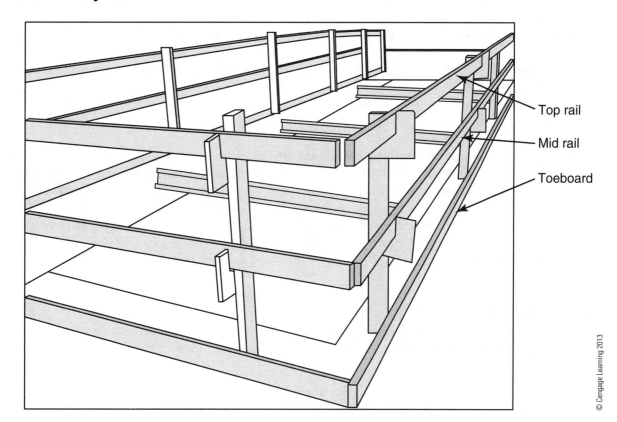

Guardrail systems

© Cengage Learning 2013

If the employer chooses to use guardrail systems to protect workers from falls, the systems must meet the following criteria. Top rails and midrails of guardrail systems must be at least ¼-inch nominal diameter or thickness to prevent cuts and lacerations. If wire rope is used for top rails, it must be flagged at not more 6-foot intervals with high-visibility material. Steel and plastic banding cannot be used as top rails or midrails. Manila, plastic, or synthetic rope used for top rails or midrails must be inspected as frequently as necessary to ensure strength and stability.

The top edge height of top rails, or (equivalent) guardrails must be 42 inches plus or minus 3 inches above the walking/working level. When workers are using stilts, the top edge height of the top rail, or equivalent member, must be increased an amount equal to the height of the stilts.

Screens, midrails, mesh, intermediate vertical members, or equivalent intermediate structural members must be installed between the top edge of the guardrail system and the walking/working surface when there are no walls or parapet walls at least 21 inches high. When midrails are used, they must be installed at a height midway between the top edge of the guardrail system and the walking/working level. When screens and mesh are used, they must extend from the top rail to the walking/working level and along the entire opening between top rail supports. Intermediate members, such as balusters, when used between posts, shall not be more than 19 inches apart.

Other structural members, such as additional midrails and architectural panels, shall be installed so that there are no openings in the guardrail system more than 19 inches.

The guardrail system must be capable of withstanding a force of at least 200 pounds applied within 2 inches of the top edge in any outward or downward direction. When the 200-pound test is applied in a downward direction, the top edge of the guardrail must not deflect to a height less than 39 inches above the walking/working level.

Midrails, screens, mesh, intermediate vertical members, solid panels, and equivalent structural members shall be capable of withstanding a force of at least 150 pounds applied in any downward or outward direction at any point along the midrails or other member.

Guardrail systems shall be surfaced to protect workers from punctures or lacerations and to prevent clothing from snagging.

The ends of top rails and midrails must not overhang terminal posts, except where such overhang does not constitute a projection hazard.

When guardrail systems are used at hoisting areas, a chain, gate, or removable guardrail section must be placed across the access opening between guardrail sections when hoisting operations are not taking place.

At holes, guardrail systems must be set up on all unprotected sides or edges. When holes are used for the passage of materials, the hole shall have not more than two sides with removable guardrail sections. When the hole is not in use, it must be covered or provided with guardrails along all unprotected sides or edges.

If guardrail systems are used around holes that are used as access points (such as ladder ways), gates must be used or the point of access must be offset to prevent accidental walking into the hole.

If guardrails are used at unprotected sides or edges of ramps and runways, they must be erected on each unprotected side or edge.

Personal Fall Arrest Systems

These consist of an anchorage, connectors, and a body harness and may include a deceleration device, lifeline, or suitable combinations. If a personal fall arrest system is used for fall protection, it must do the following:

- Limit maximum arresting force on an employee to 1,800 pounds when used with a body harness
- Be rigged so than an employee can neither free-fall more than 6 feet nor contact any lower level

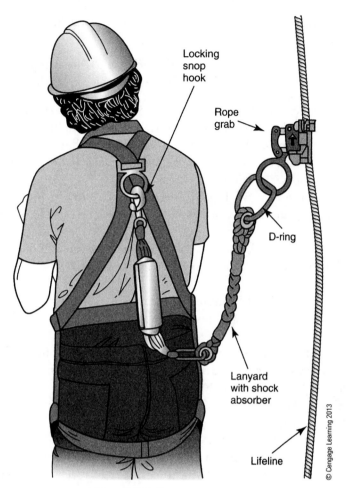

Personal fall arrest system

- Bring an employee to a complete stop and limit maximum deceleration distance an employee travels to 3.5 feet
- Have sufficient strength to withstand twice the potential impact energy of an employee free-falling a distance of 6 feet or the free-fall distance permitted by the system, whichever is less

As of January 1, 1998, the use of a body belt for fall arrest is prohibited.

Personal fall arrest systems must be inspected prior to each use for wear damage, and other deterioration. Defective components must be removed from service. D-rings and snap hooks must have a minimum tensile strength of 5,000 pounds. D-rings and snap hooks shall be proof-tested to a minimum tensile load of 3,600 pounds without cracking, breaking, or suffering permanent deformation.

Snap hooks shall be sized to be compatible with the member to which they will be connected or shall be of a locking configuration.

Unless the snap hook is a locking type and designed for the following connections, they shall not be engaged:

- Directly to webbing, rope, or wire rope
- To each other

- To a D-ring to which another snap hook or other connecter is attached
- To a horizontal lifeline
- To any object incompatible in shape or dimension relative to the snap hook, thereby causing the connected object to depress the snap hook keeper and release unintentionally.

OSHA considers a hook to be compatible when the diameter of the D-ring to which the snap hook is attached is greater than the inside length of the snap hook when measured from the bottom (hinged end) of the snap hook keeper to the inside curve of the top of the snap hook. Thus, no matter how the D-ring is positioned or moved (rolls) with the snap hook attached, the D-ring cannot touch the outside of the keeper, thus depressing it open. The use of nonlocking snap hooks is prohibited.

On suspended scaffolds or similar work platforms with horizontal lifelines that may become vertical lifelines, the devices used to connect to a horizontal lifeline shall be capable of locking in both directions on the lifeline.

Horizontal lifelines shall be designed, installed, and used under the supervision of a qualified person, as part of a complete personal fall arrest system that maintains a safety factor of at least two. Lifelines shall be protected against being cut or abraded.

Self-retracting lifelines and lanyards that automatically limit free-fall distance to 2 feet or less shall be capable of sustaining a minimum tensile load of 3,000 pounds applied to the device with the lifeline or lanyard in the fully extended position.

Self-retracting lifelines and lanyards that do not limit free-fall distance to 2 feet or less, rip stitch lanyards, and tearing and deforming lanyards shall be capable of sustaining a minimum tensile load of 5,000 pounds applied to the device with the lifeline or lanyard in the fully extended position.

Ropes and straps (webbing) used in lanyards, lifelines, and strength components of body belts and body harnesses shall be made of synthetic fibers.

Anchorages shall be designed, installed, and used under the supervision of a qualified person, as part of a complete personal fall arrest system that maintains a safety factor of at least two, i.e., capable of supporting at least twice the weight expected to be imposed upon it. Anchorages used to attach personal fall arrest systems shall be independent of any anchorage being used to support or suspend platforms and must be capable of supporting at least 5,000 pounds per person attached. Lanyards and vertical lifelines must have a minimum breaking strength of 5,000 pounds.

Positioning Device Systems

These body belt or body harness systems are to be set up so that a worker can free-fall no farther than 2 feet. They shall be secured to an anchorage capable of supporting at least twice the potential impact load of an employee's fall or 3,000 pounds, whichever is greater. Requirements for snap hooks, D-rings, and other connectors used with positioning device systems must meet the same criteria as those for personal fall arrest systems.

Safety Monitoring Systems

When no other alternative fall protection has been implemented, the employer shall implement a safety monitoring system. Employers must appoint a competent person to monitor the safety of workers and the employer shall ensure that the safety monitor:

- Is competent in the recognition of fall hazards
- Is capable of warning workers of fall hazard dangers and in detecting unsafe work practices
- Is operating on the same walking/working surfaces as the workers and can see them
- Is close enough to work operations to communicate orally with workers and has no other duties to distract from the monitoring function

Mechanical equipment shall not be used or stored in areas where safety monitoring systems are being used to monitor employees engaged in roofing operations on low-sloped roofs.

No worker, other than one engaged in roofing work (on low-sloped roofs) or one covered by a fall protection plan, shall be allowed in an area where an employee is being protected by a safety monitoring system.

All workers in a controlled access zone shall be instructed to promptly comply with fall hazard warnings issued by safety monitors.

Safety Net Systems

Safety nets must be installed as close as practicable under the walking/working surface on which employees are working and never more than 30 feet below such levels. Defective nets shall not be used. Safety nets shall be inspected at least once a week for wear, damage, and other deterioration. The maximum size of each safety net mesh opening shall not exceed 36 square inches nor be longer than 6 inches on any side, and the openings, measured center-to-center, of mesh ropes or webbing shall not exceed 6 inches. All mesh crossings shall be secured to prevent enlargement of the mesh opening. Each safety net or section shall have a border rope for webbing with a minimum breaking strength of 5,000 pounds. Connections between safety net panels shall be as strong as integral net components and be spaced no more than 6 inches apart.

Safety nets shall be installed with sufficient clearance underneath to prevent contact with the surface or structure below.

When nets are used on bridges, the potential fall area from the walking/working surface to the net shall be unobstructed.

Safety nets must extend outward from the outermost projection of the work surface as follows:

Vertical Distance from Working Level to Horizontal Plane of Net	Minimum Required Horizontal Distance of Outer Edge of Net from the Edge of the Working Surface
Up to 5 feet (1.5 meters)	8 feet (2.4 meters)
More than 5 feet (1.5 meters) up to 10 feet (3 meters)	10 feet (3 meters)
More than 10 feet (3 meters)	13 feet (3.9 meters)

Safety nets shall be capable of absorbing an impact force of a drop test consisting of a 400-pound bag of sand 30 inches in diameter dropped from the highest walking/working surface at which workers are exposed, but not from less than 42 inches above that level.

Items that have fallen into safety nets including—but not restricted to—materials, scrap, equipment, and tools must be removed as soon as possible and at least before the next work shift.

Warning-Line Systems

Warning-line systems consist of ropes, wires, or chains and supporting stanchions and are set up as follows:

- Flagged at not more than 6-foot intervals with high-visibility material
- Rigged and supported so that the lowest point (including sag) is no less than 34 inches from the walking/working surface and its highest point is no more than 39 inches from the walking/working surface

- Stanchions, after being rigged with warning lines, shall be capable of resisting, without tipping over, a force of at least 16 pounds applied horizontally against the stanchion, 30 inches above the walking/working surface, perpendicular to the warning line and in the direction of the floor, roof, or platform edge

- The rope, wire, or chain shall have a minimum tensile strength of 500 pounds and after being attached to the stanchions must support, without breaking, the load applied to the stanchions as prescribed above

- The rope, wire, or chain shall be attached to each stanchion in such a way that pulling on one section of the line between stanchions will not result in slack being taken up in the adjacent section before the stanchion tips over

Warning lines shall be erected around all sides of roof work areas. When mechanical equipment is being used, the warning line shall be erected not less than 6 feet from the roof edge parallel to the direction of mechanical equipment operation, and not less than 10 feet from the roof edge perpendicular to the direction of mechanical equipment operation.

When mechanical equipment is not being used, the warning line must be erected not less than 6 feet from the roof edge.

Covers

Covers located in roadways and vehicular aisles must be able to support at least twice the maximum axle load of the largest vehicle to which the cover might be subjected. All other covers must be able to support at least twice the weight of employees, equipment, and materials that may be imposed on the cover at any one time. To prevent accidental displacement resulting from wind, equipment, or workers' activities, all covers must be secured. All covers shall be color coded or bear the markings "HOLE" or "COVER."

Protection from Falling Objects

When guardrail systems are used to prevent materials from falling from one level to another, any openings must be small enough to prevent passage of potential falling objects. No materials or equipment except masonry and mortar shall be stored within 4 feet of working edges. Excess mortar, broken or scattered masonry units, and all other materials and debris shall be kept clear of the working area by removal at regular intervals.

During roofing work, materials and equipment shall not be stored within 6 feet of a roof edge unless guardrails are erected at the edge, and materials piled, grouped, or stacked near a roof edge must be stable and self-supporting.

Canopies

When used as protection from falling objects canopies must be strong enough to prevent collapse and to prevent penetration by any objects that may fall onto them.

Toe Boards

When toe boards are used as protection from falling objects, they must be erected along the edges of the overhead walking/working surface for a distance sufficient to protect persons working below. Toe boards shall be capable of withstanding a force of at least 50 pounds applied in any downward or outward direction at any point along the toe board.

Toe boards shall be a minimum of 3.5 inches tall from their top edge to the level of the walking/working surface, have no more than 0.25-inch clearance above the walking/working surface, and be solid or have openings no larger than 1 inch in size. Where tools, equipment, or materials are piled higher than the top edge of a toe board, paneling or screening must be erected from the walking/working surface or toe board to the top of a guardrail system's top rail or midrail for a distance sufficient to protect employees below.

Fall Protection Training

Employers must provide a training program that teaches employees who might be exposed to fall hazards how to recognize such hazards and how to minimize them. Employees must be trained in the following areas:

- The nature of fall hazards in the work area
- The correct procedures for erecting, maintaining, disassembling, and inspecting fall protection systems
- The use and operation of controlled access zones and guardrail, personal fall arrest, safety net, warning-line, and safety monitoring systems
- The role of each employee in the safety monitoring system when the system is in use
- The limitations on the use of mechanical equipment during the performance of roofing work on low-sloped roofs
- The correct procedures for equipment and materials handling and storage and the erection of overhead protection
- Employees' role in fall protection plans

Employers must prepare a written certification that identifies the employee trained and the date of the training. The employer or trainer must sign the certification record. Retraining also must be provided when necessary.

CRANES AND DERRICKS IN CONSTRUCTION

In 2010, OSHA created a new subpart to address crane and derrick safety in construction. These new regulations can be found in Subpart CC.

Assembly/Disassembly Procedures

Under this standard, employers must comply with all manufacturer prohibitions regarding assembly and disassembly. However, the standard generally allows employers to choose between the manufacturer's procedures or their own (see exception below for synthetic sling procedures). Employer procedures must be developed by a "qualified person" and must satisfy a number of specified requirements, such as providing adequate support and stability for all parts of the equipment, and positioning employees involved to minimize exposure to any unintended movement or collapse.

Assembly/Disassembly Responsibilities

- The rule requires the work to be directed by an A/D (Assembly/Disassembly) director. The A/D director must meet the criteria for both a "competent person" and a "qualified person," which are defined terms in this rule, or must be a "competent person" assisted by a "qualified person."
- The A/D director must understand the applicable procedures.
- The A/D director must review the procedures immediately prior to beginning work unless he or she understands the procedures and has used them before for that equipment type and configuration.
- The A/D director must ensure that each member of the crew understands his or her tasks, the hazards of the tasks, and any hazardous positions or locations to avoid.
- The A/D director must verify all capacities of any equipment used, including rigging, lifting lugs, etc.
- The A/D director must also address hazards associated with the operation, including 12 specified areas of concern: site and ground conditions, blocking material, proper location of blocking, verifying

assist crane loads, boom and jib pick points, center of gravity, stability upon pin removal, snagging, struck by counterweights, boom hoist brake failure, loss of backward stability, and wind speed and weather.

Inspection

- Upon completion of assembly, but before use, the equipment must be inspected by a "qualified person" to ensure that it is configured in accordance with the manufacturer equipment criteria. If these criteria are unavailable, the employer's "qualified person," with the assistance of a registered professional engineer if necessary, must develop the appropriate configuration criteria and ensure that these criteria are met.

General Requirements

- A crew member who moves out of the operator's view to a location where the crew member could be injured by movement of the equipment (or load) MUST inform the operator before going to that location. The operator must not move the equipment until that crew member informs the operator that he or she has relocated to a safe position.
- Employees must never be under the boom or jib when pins (or similar devices) are being removed, unless it is required by site constraints and the A/D director has implemented procedures that minimize the risk of unintended movement and the duration and extent of exposure under the boom.
- Component weights must be readily available for all components to be assembled.
- All rigging must be done by a "qualified rigger."
- Pins may not be removed during disassembly when the pendants are in tension.
- Booms supported only by cantilevering must not exceed manufacturer limitations or RPE limitations, as applicable.
- Component selection and equipment configuration that affects the capacity or safe operation of the equipment must be in accordance with manufacturer requirements and limits or RPE requirements and limits, as applicable.

Synthetic Slings

- The employer must follow manufacturer procedures when using synthetic slings during assembly or disassembly of rigging (even when the employer has developed its own A/D procedure as an alternative to the manufacturer's other procedures.)
- Synthetic slings must be protected from abrasive, sharp, or acute edges and configurations that might reduce the sling's rated capacity.

Outriggers and Stabilizers

When outriggers or stabilizers are used or are necessary in light of the load to be handled and the operating radius:

- Outriggers and stabilizers must be fully extended or, if permitted by manufacturer procedures, deployed as specified in the load chart.
- Outriggers must be set to remove equipment weight from the wheels, except for locomotive cranes.
- Outrigger floats, if used, must be attached to the outriggers; stabilizer floats, if used, must be attached to the stabilizers.

- Each outrigger or stabilizer must be visible to the operator or to a signal person during extension and setting.
- Outrigger and stabilizer blocking must be placed under the float/pad of the jack or, if there is no jack, under the outer bearing surface of the outrigger or stabilizer beam. Blocking must also be sufficient to sustain the loads and maintain stability and must be properly placed.

Tower Cranes

- Tower cranes are subject to additional requirements for erecting, climbing, and dismantling, including a pre-erection inspection (29 CFR 1926.1435).

Operator Qualification and Certification

Who Needs to Be Certified or Qualified?

Any person engaged in a construction activity that is operating a crane covered by the new cranes and derricks rule, except:

- Sideboom cranes
- Derricks
- Equipment with a rated hoisting/lifting capacity of 2,000 pounds or less

NOTE: Operators of the equipment listed above must meet the criteria for minimum expertise described in the applicable section in subpart CC.

Are Operators of Digger Derricks Required to Be Qualified or Certified?

Yes, unless the digger derrick is being used to auger holes for poles carrying electric or telecommunication lines, place or remove the poles, or handle associated materials to be installed on or removed from the poles.

What Is Required in the Testing for Certification?

Certification has two parts:

1. A written examination that includes the safe operating procedures for the particular type of equipment the applicant will be operating and technical understanding of the subject matter criteria
2. A practical exam showing the applicant has the skills needed to safely operate the equipment, including, among other skills, the ability to properly use load chart information and recognize items required in the shift inspection

Does an Operator Need More Than One Certification?

With respect to certification from an accredited testing organization, an operator must be certified for the type and capacity of crane he or she is going to operate. Each accredited testing organization develops its own categories for crane type and capacity.

How Is an Operator Certified or Qualified?

There are four ways that an equipment operator can be qualified or certified and meet OSHA requirements.

1. A certificate from an accredited crane operator testing organization
2. Qualification from the employer through an audited employer program
3. Qualification by the U.S. military (only applies to employees of the Department of Defense or Armed Forces and does not include private contractors)

4. Licensing by a state or local government (if that licensing meets the minimum requirements set forth by OSHA) (NOTE: When a state or local government requires a crane operator license, the crane operator must be licensed accordingly to meet OSHA requirements.)

Accredited Crane Operator Testing Organization

The testing organization must be accredited by a nationally recognized accrediting agency and test according to the criteria listed at 1926.1427(j)(1) and (j)(2). This certification is portable from employer to employer. The testing organization must have its accreditation reviewed every 3 years. The certificate must note the type and capacity of equipment for which the operator is tested and certified. The certificate is valid for 5 years.

Audited Employer Program

An employer may provide a crane operator testing program under the oversight of an independent auditor.

An accredited crane operator testing organization must certify the auditor to evaluate the administration of written and practical tests. The auditor must conduct audits of the employer's program according to nationally recognized auditing standards. Crane operator qualification under an employer program is only valid while the operator is an employee of the employer and operating a crane for the employer. The qualification is valid up to 5 years.

U.S. Military

This qualification applies only to civilian employees of the Department of Defense or Armed Services and is not portable. This qualification does not include employees of private contractors.

Licensing by a Government Entity

This license is obtained from a government entity, such as a city or state that has a required certification program. When this license meets the minimum requirements of 1926.1427(e)(2) and (j), OSHA requires a crane operator to have this license when operating in the applicable city, county, or state. This license is not portable outside the boundaries of the government entity that issues the license and is valid for a maximum of 5 years.

Qualified Rigger

When Is a Qualified Rigger Required?

Employers must use qualified riggers during hoisting activities for assembly and disassembly work. Additionally, qualified riggers are required whenever workers are within the fall zone and hooking, unhooking, or guiding a load, or doing the initial connection of a load to a component or structure.

Who Can Be a Qualified Rigger?

A qualified rigger is a rigger who meets the criteria for a qualified person. Employers must determine whether a person is qualified to perform specific rigging tasks. Each qualified rigger may have different credentials or experience. A qualified rigger is a person that:

- Possesses a recognized degree, certificate, or professional standing, or
- Has extensive knowledge, training, and experience, and
- Can successfully demonstrate the ability to solve problems related to rigging loads.

The person designated as the qualified rigger must have the ability to properly rig the load for a particular job. It does not mean that a rigger must be qualified to do every type of rigging job.

Each load that requires rigging has unique properties that can range from the simple to the complex. For example, a rigger may have extensive experience in rigging structural components and other equipment to support specific construction activities. Such experience may have been gained over many years. However, this experience does not automatically qualify the rigger to rig unstable, unusually heavy, or eccentric loads that may require a tandem lift, multiple lifts, or use of custom rigging equipment. In essence, employers must make sure the person can do the rigging work needed for the exact types of loads and lifts for a particular job with the equipment and rigging that will be used for that job.

Do Qualified Riggers Have to Be Trained or Certified by an Accredited Organization or Assessed by a Third Party?

No. Riggers do not have to be certified by an accredited organization or assessed by a third party. Employers may choose to use a third-party entity to assess the qualifications of the rigger candidate, but they are not required to do so.

Does a Certified Operator Also Meet the Requirements of a Qualified Rigger?

A certified operator does not necessarily meet the requirements of a qualified rigger. Determining whether a person is a qualified rigger is based on the nature of the load, lift, and equipment used to hoist that load plus that person's knowledge and experience. A certified/qualified operator may meet the requirements of a qualified rigger, depending on the operator's knowledge and experience with rigging.

Signal Person Qualification

When Is a Signal Person Required?

A signal person is required when:

- The point of operation is not in full view of the operator.
- The operator's view is obstructed in the direction the equipment is traveling.
- Either the operator or the person handling the load determines that a signal person is needed because of site-specific safety concerns.

What Does a Signal Person Need to Know?

The signal person is considered qualified if he or she:

- Knows and understands the type of signals used at the worksite
- Is competent in using these signals
- Understands the operations and limitations of the equipment, including the crane dynamics involved in swinging, raising, lowering, and stopping loads and in boom deflection from hoisting loads
- Knows and understands the relevant signal person qualification requirements specified in subpart CC
- Passes an oral or written test and a practical test

How Does a Signal Person Become Qualified?

Employers must use one of the following options to ensure that a signal person is qualified:

1. Third-party qualified evaluator. The signal person has documentation from a third-party qualified evaluator showing that he or she meets the qualification requirements
2. Employer's qualified evaluator (not a third party). The employer's qualified evaluator assesses the individual, determines the individual meets the qualification requirements, and provides documentation of that determination. This assessment may not be relied on by other employers

How Will an Employer Show That a Signal Person Is Appropriately Qualified?

Employers must make the documentation of the signal person's qualifications available at the worksite, either in paper form or electronically. The documentation must specify each type of signaling (e.g., hand signals, radio signals, etc.) for which the signal person is qualified under the requirements of the standard.

MOTOR VEHICLES AND MECHANIZED EQUIPMENT

General Requirements

All vehicles in use shall be checked at the beginning of each shift to ensure that all parts, equipment, and accessories that affect safe operation are in proper operating condition and free from defects. All defects shall be corrected before the vehicle is placed in service.

No employer shall use any motor vehicle, earthmoving, or compacting equipment having an obstructed view to the rear unless:

- The vehicle has a reverse signal alarm distinguishable from the surrounding noise level, or
- The vehicle is backed up only when an observer signals that it is safe to do so.

Heavy machinery, equipment, or parts thereof that are suspended or held aloft shall be substantially blocked to prevent falling or shifting before employees are permitted to work under or between them.

Safety Driving Practices

Construction companies should develop driver safety rules for every worker. Here are some basic rules to share with your workers:

- Stay Safe
  - Use a seat belt at all times—driver and passenger(s).
  - Be well rested before driving.
  - Avoid taking medications that make you drowsy.
  - Set a realistic goal for the number of miles that you can drive safely each day.
  - If you are impaired by alcohol or any drug, do not drive.
- Stay Focused
  - Driving requires your full attention. Avoid distractions, such as adjusting the radio or other controls, eating or drinking, and talking on the phone.
  - Continually search the roadway to be alert to situations requiring quick action.
  - Stop about every 2 hours for a break. Get out of the vehicle to stretch, take a walk, and get refreshed.
- Avoid Aggressive Driving
  - Keep your cool in traffic!
  - Be patient and courteous to other drivers.
  - Do not take other drivers' actions personally.
  - Reduce your stress by planning your route ahead of time (bring the maps and directions), allowing plenty of travel time, and avoiding crowded roadways and busy driving times.

Distracted Driving

Distracted driving has many dangers:

- Distracted driving crashes killed more than 5,400 people and injured nearly 500,000 in 2009.
- Researchers report that texting while driving claimed more than 16,000 lives from 2001 to 2007.
- Reaction time is delayed for a driver talking on a cell phone as much as it is for a driver who is legally drunk.
- Drivers who are texting take their eyes off the road 400% more than when they are not texting.
- More texting leads to more crashes. With each additional 1 million text messages, fatalities from distracted driving rose more than 75%.

To fight these dangers, construction companies should create rules to prevent distracted driving. Here are some suggested rules:

- Prohibit texting while driving. OSHA encourages employers to declare their vehicles "text-free zones" and to emphasize that commitment to their workers, customers, and communities.
- Establish work procedures and rules that do not make it necessary for workers to text while driving in order to carry out their duties.
- Set up clear procedures, times, and places for drivers' safe use of texting and other technologies for communicating with managers, customers, and others.
- Incorporate safe communications practices into worker orientation and training.
- Eliminate financial and other incentive systems that encourage workers to text while driving.

EXCAVATIONS

OSHA issued its first Excavation and Trenching Standard in 1971 to protect workers from excavation hazards. Since then, OSHA has amended the standard several times to increase worker protection and to reduce the frequency and severity of excavation accidents and injuries. Despite these efforts, excavation-related accidents resulting in injuries and fatalities continue to occur.

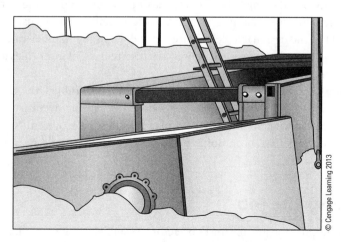

Trench with cave-in protection

The following are highlights from the regulations. Construction companies who deal with excavation and trenching operations will need to follow these safe work practices.

According to the OSHA construction safety and health standards, a trench is referred to as a narrow excavation made below the surface of the ground in which the depth is greater than the width—the width not exceeding 15 feet. An excavation is any man-made cut, cavity, trench, or depression in the earth's surface formed by earth removal. This can include excavations for anything from cellars to highways.

Planning

Many on-the-job accidents are a direct result of inadequate initial planning. Correcting mistakes in shoring and/or sloping after work has begun slows down the operation, adds to the cost, and increases the possibility of an excavation failure. The contractor should build safety into the pre-bid planning in the same way all other pre-bid factors are considered.

It is a good idea for contractors to develop safety checklists before preparing a bid, to make certain there is adequate information about the job and all needed items are on hand. These checklists should incorporate elements of the relevant OSHA standards as well as other information necessary for safe operations.

Before preparing a bid, these specific site conditions should be taken into account:

- Traffic
- Nearness of structures and their conditions
- Soil
- Surface and ground water
- The water table
- Overhead and underground utilities
- Weather

These and other conditions can be determined by jobsite studies, observations, test borings for soil type or conditions, and consultations with local officials and utility companies.

Before any excavation actually begins, the standard requires the employer to determine the estimated location of utility installations—sewer, telephone, fuel, electric, and water lines, or any other underground installations—that may be encountered during digging. Also, before starting the excavation, the contractor must contact the utility companies or owners involved and inform them, within established or customary local response times, of the proposed work. 811 is the designated "call before you dig" phone number throughout the United States that directly connects you to your local one-call center. Each state has different rules and regulations governing digging, some stricter than others.

When all the necessary specific information about the job site is assembled, the contractor is ready to determine the amount, kind, and cost of the safety equipment needed. A careful inventory of the safety items on hand should be made before deciding what additional safety material must be acquired. No matter how many trenching, shoring, and backfilling jobs have been done in the past, each job should be approached with the utmost care and preparation.

Before Beginning the Job

It is important, before beginning the job, for the contractor to establish and maintain a safety and health program for the worksite that provides adequate systematic policies, procedures, and practices to protect employees from, and allow them to recognize, job-related safety and health hazards.

It is also important, before beginning work, for employers to provide employees who are exposed to public vehicular traffic with warning vests or other suitable garments marked with or made of reflectorized or high-visibility material and ensure that they wear them. Workers must also be instructed to remove or neutralize surface encumbrances that may create a hazard.

In addition, no employee should operate a piece of equipment without first being properly trained to handle it and fully alerted to its potential hazards.

In the training and in the site safety and health program, it also is important to incorporate procedures for fast notification and investigation of accidents.

On-the-Job-Evaluation

The standard requires that a competent person inspect, on a daily basis, excavations and the adjacent areas for possible cave-ins, failures of protective systems and equipment, hazardous atmospheres, or other hazardous conditions. If these conditions are encountered, exposed employees must be removed from the hazardous area until the necessary safety precautions have been taken. Inspections are also required after natural (e.g., heavy rains) or man-made events such as blasting that may increase the potential for hazards.

Larger and more complex operations should have a full-time safety official who makes recommendations to improve the implementation of the safety plan. In a smaller operation, the safety official may be part-time and usually will be a supervisor.

Supervisors are the contractor's representatives on the job. Supervisors should conduct inspections, investigate accidents, and anticipate hazards. They should ensure that employees receive on-the-job safety and health training. They should also review and strengthen overall safety and health precautions to guard against potential hazards, get the necessary worker cooperation in safety matters, and make frequent reports to the contractor.

It is important that managers and supervisors set the example for safety at the jobsite. It is essential that when visiting the jobsite, all managers, regardless of status, wear the prescribed personal protective equipment such as safety shoes, safety glasses, hard hats, and other necessary gear.

Employees must also take an active role in job safety. The contractor and supervisor should make certain that workers have been properly trained in the use and fit of the prescribed protective gear and equipment, that they are wearing and using the equipment correctly, and that they are using safe work practices.

Mechanics of a Cave-in

An open excavation is an unnatural situation. The average landscape shows no vertical or near-vertical slopes.

Soil or dirt is a very heavy material. A cubic foot can easily weigh 114 pounds and a cubic yard can be as heavy as a pick-up truck.

Consider a column of soil which is 1 foot by 1 foot and several feet high. At 1 foot of depth a horizontal plane 1 foot by 1 foot is carrying the cubic foot which lies over it. The stress, or load per unit area, is 100 lb. per square foot (psf). At a depth of 2 feet the horizontal plane is carrying 2 cubic feet, or 200 psf. At a depth of 5 feet the vertical stress is 500 psf, and so on.

The column described would soon collapse if not supported by similar adjacent columns. Stresses are developed that act horizontally on the column. These lateral stresses can be considered to be one-half as large as the vertical stresses. At a depth of 5 feet the vertical stress is 500 psf and the lateral, or horizontal, stress is 250 psf. Undisturbed soil may be visualized as an infinite number of columns of soil adjoining and supporting one another. The system is in equilibrium and is perfectly stable.

When an excavation is cut the system is disturbed. Lateral stresses that existed on the excavation wall are removed as the excavation is done. The soil in the excavation wall immediately begins to move, however slowly, into the excavation.

At the same time, the surface of the ground next to the excavation subsides, creating an unnatural situation. The surface of the ground is in tension and some of the weight of the soil in the excavation wall is transferred to the soil back away from the wall face by a phenomenon called shear.

The combination of tension in the ground surface and shear stress causes cracks to form back from the edge of the excavation. Cracks occur to the depth of the excavation back from its edge. For example, if an excavation 10 feet deep is dug, the cracks may be found somewhere between 3 to 7 feet back from the excavation edge. There may be several cracks. They are usually vertical and they may be one-half the depth of the excavation.

When cracks develop, the weight of the soil in the excavation wall is no longer partly carried by the soil back from the excavation's face. Then the lower part of the excavation wall fails under the great stress from the weight of the soil above it. There is no lateral stress to prevent the failure.

When the bottom of the excavation fails, or "kicks," into the excavation, the support for the upper part of the excavation wall is now essentially hanging only by shear and tension forces. Failure occurs.

A third cave-in quickly follows. Soil, like concrete, is normally strong in compression but not at all strong in tension. Reinforced concrete makes use of the compressive strength of concrete and the tensile strength of steel. There is no steel in the soil.

Cave-ins generally come in multiples. If the first one doesn't get you, the second one may and the third is always a possibility.

This example of the mechanics of a cave-in has offered a discussion of some of the forces involved in such accidents. It has by no means considered all of the forces that may be involved in such an occurrence. Weathering, water, vibration, and superimposed loads may add to the hazardous conditions leading to cave-ins.

Protective Support Systems

Excavation workers are exposed to many hazards, but the chief hazard is danger of cave-ins. OSHA requires that in all excavations employees exposed to potential cave-ins must be protected by sloping or benching the sides of the excavation; supporting the sides of the excavation; or placing a shield between the side of the excavation and the work area. Designing a protective system can be complex because of the number of factors involved—soil classification, depth of cut, water content of soil, changes due to weather and climate, or other operations in the vicinity. The standard, however, provides several different methods and approaches (four for sloping and four for shoring, including the use of shields) for designing protective systems that can be used to provide the required level of protection against cave-ins.

One method of ensuring the safety and health of workers in an excavation is to slope the sides to an angle not steeper than 1½ horizontal to 1 vertical (34 degrees measured from the horizontal). A slope of this gradation or less is considered safe for any type of soil. A field method for identifying soil can be found in Resource L, located at the back of this book and at dewalt.com/guides.

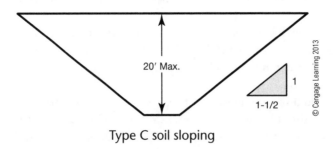

20' Max.

1

1-1/2

© Cengage Learning 2013

Type C soil sloping

A second design method, which can be applied for both sloping and shoring, involves using tabulated data, such as tables and charts, approved by a registered professional engineer. These data must be in writing and must include sufficient explanatory information to enable the user to make a selection, including the criteria for determining the selection and the limits on the use of the data.

At least one copy of the information, including the identity of the registered professional engineer who approved the data, must be kept at the worksite during construction of the protective system. Upon completion of the system, the data may be stored away from the jobsite, but a copy must be made available, upon request, to the Assistant Secretary of Labor for OSHA.

Contractors also may use a trench box or shield that is either designed or approved by a registered professional engineer or is based on tabulated data prepared or approved by a registered professional engineer. Timber, aluminum, or other suitable materials may also be used. OSHA standards permit the use of a trench box or shield as long as the protection it provides is equal to or greater than the protection that would be provided by the appropriate shoring system.

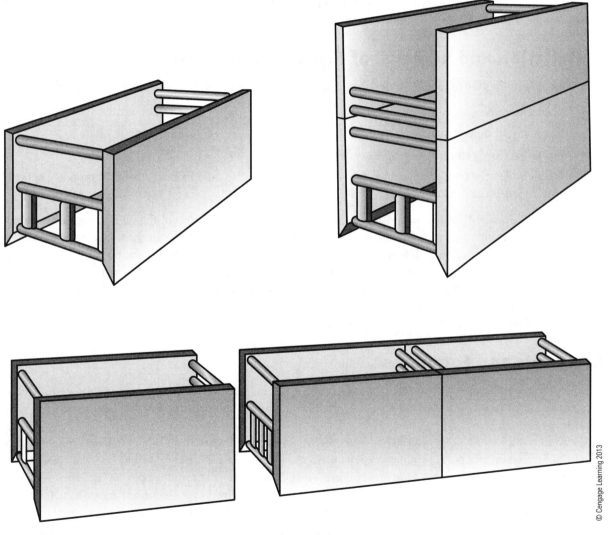

Types of trench boxes

The employer is free to choose the most practical design approach for any particular circumstance. Once an approach has been selected, however, the required performance criteria must be met by that system.

The standard does not require the installation and use of a protective system when an excavation (1) is made entirely in stable rock or (2) is less than 5 feet deep and a competent person has examined the ground and found no indication of a potential cave-in.

Safety Precautions

The standard requires the employer to provide support systems such as shoring, bracing, or underpinning to ensure the stability of adjacent structures such as buildings, walls, sidewalks or pavements.

The standard prohibits excavation below the level of the base or footing of any foundation or retaining wall unless:

1. A support system such as underpinning is provided
2. The excavation is in stable rock
3. A registered professional engineer determines that the structure is sufficiently removed from the excavation and that excavation will not pose a hazard to employees

Excavations under sidewalks and pavements are also prohibited unless an appropriately designed support system is provided or another effective method is used.

Installation and Removal of Protective Systems

The standard requires the following procedures for the protection of employees when installing support systems:

- Securely connect members of support systems
- Safely install support systems
- Never overload members of support systems
- Install other structural members to carry loads imposed on the support system when temporary removal of individual members is necessary

In addition, the standard permits excavation of 2 feet or less below the bottom of the members of a support or shield system of a trench if (1) the system is designed to resist the forces calculated for the full depth of the trench and (2) there are no indications, while the trench is open, of a possible cave-in below the bottom of the support system. Also, the installation of support systems must be closely coordinated with the excavation of trenches.

As soon as work is completed, the excavation should be backfilled as the protective system is dismantled. After the excavation has been cleared, workers should slowly remove the protective system from the bottom up, taking care to release members slowly.

Materials and Equipment

The employer is responsible for the safe condition of materials and equipment used for protective systems. Defective and damaged materials and equipment can result in the failure of a protective system and cause excavation hazards.

To avoid possible failure of a protective system, the employer must ensure that (1) materials and equipment are free from damage or defects, (2) manufactured materials and equipment are used and maintained in a manner consistent with the recommendations of the manufacturer and in a way that will prevent employee exposure to hazards, and (3) while in operation, damaged materials and equipment are examined by a competent person to determine if they are suitable for continued use. If materials and equipment are not safe for use, they must be removed from service. These materials cannot be returned to service without the evaluation and approval of a registered professional engineer.

Other Hazards

Falls and Equipment

In addition to cave-in hazards and secondary hazards related to cave-ins, there are other hazards from which workers must be protected during excavation-related work. These hazards include exposure to falls, falling loads, and mobile equipment. To protect employees from these hazards, OSHA requires the employer to take the following precautions:

- Keep materials or equipment that might fall or roll into an excavation at least 2 feet from the edge of excavations, or have retaining devices, or both

- Provide warning systems such as mobile equipment, barricades, hand or mechanical signals, or stop logs to alert operators of the edge of an excavation. If possible, keep the grade away from the excavation

- Provide scaling to remove loose rock or soil or install protective barricades and other equivalent protection to protect employees against falling rock, soil, or materials

- Prohibit employees from working on faces of sloped or benched excavations at levels above other employees unless employees at lower levels are adequately protected from the hazard of falling, rolling, or sliding material or equipment

- Prohibit employees under loads that are handled by lifting or digging equipment. To avoid being struck by any spillage or falling materials, require employees to stand away from vehicles being loaded or unloaded. If cabs of vehicles provide adequate protection from falling loads during loading and unloading operations, the operators may remain in them

Water Accumulation

The standard prohibits employees from working in excavations where water has accumulated or is accumulating unless adequate protection has been taken. If water-removal equipment is used to control or prevent water from accumulating, the equipment and operations of the equipment must be monitored by a competent person to ensure proper use.

OSHA standards also require that diversion ditches, dikes, or other suitable means be used to prevent surface water from entering an excavation and to provide adequate drainage of the area adjacent to the excavation. Also, a competent person must inspect excavations subject to runoffs from heavy rains.

Hazardous Atmospheres

Under this provision, a competent person must test excavations greater than 4 feet in depth as well as ones where oxygen deficiency or a hazardous atmosphere exists or could reasonably be expected to exist before an employee enters the excavation. If hazardous conditions exist, controls such as proper respiratory protection or ventilation must be provided. Also, controls used to reduce atmospheric contaminants to acceptable levels must be tested regularly.

Where adverse atmospheric conditions may exist or develop in an excavation, the employer also must provide and ensure that emergency rescue equipment, (e.g., breathing apparatus, a safety harness and line, basket stretcher, etc.) is readily available. This equipment must be attended when used.

When an employee enters bell-bottom pier holes and similar deep and confined footing excavations, the employee must wear a harness with a lifeline. The lifeline must be securely attached to the harness and must be separate from any line used to handle materials. Also, while the employee wearing the lifeline is in the excavation, an observer must be present to ensure that the lifeline is working properly and to maintain communication with the employee.

Access and Egress

Under the standard, the employer must provide safe access and egress to all excavations. According to OSHA regulations, when employees are required to be in trench excavations 4 feet deep or more, adequate means of exit, such as ladders, steps, ramps, or other safe means of egress, must be provided and be within 25 feet of lateral travel. If structural ramps are used as a means of access or egress, they must be designed by a competent person if used for employee access or egress or a competent person qualified in structural design if used by vehicles. Also, structural members used for ramps or runways must be uniform in thickness and joined in a manner to prevent tripping or displacement.

Summary

Trenching and excavation work presents serious risks to all workers involved. The greatest risk, and one of primary concern, is that of a cave-in. Furthermore, when cave-in accidents occur, they are much more likely to result in worker fatalities than other excavation-related accidents. Strict compliance, however, with all sections of the standard will prevent or greatly reduce the risk of cave-ins as well as other excavation-related accidents.

CONCRETE AND MASONRY CONSTRUCTION

Construction companies must comply with the OSHA regulations on concrete and masonry to protect construction workers from accidents and injuries resulting from the premature removal of formwork, the failure to brace masonry walls, the failure to support precast panels, the inadvertent operation of equipment, and the failure to guard reinforcing steel.

The OSHA regulation, found in Subpart Q, prescribes performance-oriented requirements designed to help protect all construction workers from the hazards associated with concrete and masonry construction operations at construction, demolition, alteration, or repair worksites.

General Requirements

Construction Loads

Employers must not place construction loads on a concrete structure or portion of a concrete structure unless the employer determines, based on information received from a person who is qualified in structural design, that the structure or portion of the structure is capable of supporting the intended loads.

Reinforcing Steel

All protruding reinforcing steel, onto and into which employees could fall, must be guarded to eliminate the hazard of impalement. The first line of defense would be making sure the employees cannot fall on the rebar. Rebar caps or bending over the rebar is another option. Note: Mushroom-style rebar caps will not provide impalement protection.

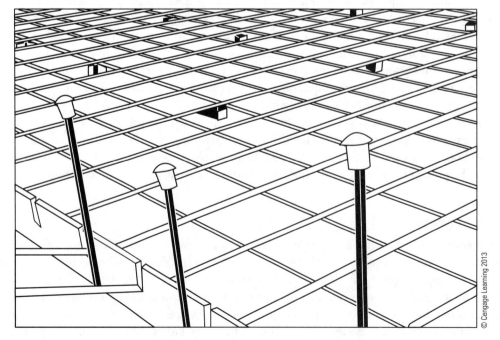

Mushroom cap—improper impalement protection

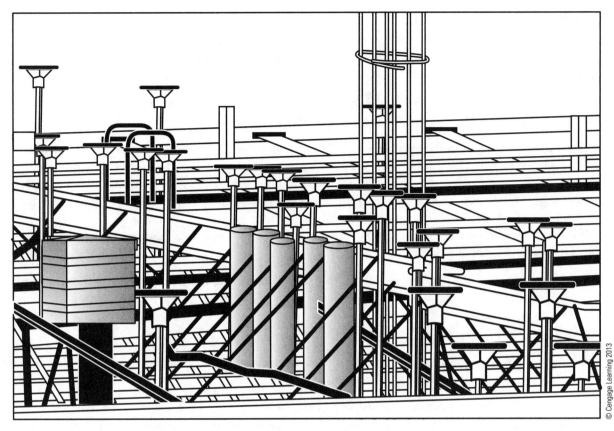

Proper rebar impalement protection

Post-Tensioning Operations

Employees (except those essential to the post-tensioning operations) must not be permitted to be behind the jack during tensioning operations.

Signs and barriers must be erected to limit employee access to the post-tensioning area during tensioning operations.

Concrete Buckets

Employees must not be permitted to ride concrete buckets.

Working Under Loads

Employees must not be permitted to work under concrete buckets while the buckets are being elevated or lowered into position.

To the extent practicable, elevated concrete buckets must be routed so that no employee or the fewest employees possible are exposed to the hazards associated with falling concrete buckets.

Personal Protective Equipment

Employees must not be permitted to apply a cement, sand, and water mixture through a pneumatic hose unless they are wearing protective head and face equipment. Portland cement contains hexavalent chromium, which can cause serious skin disorders, and workers should use appropriate gloves, boots, and other PPE to prevent direct contact with their skin. Employees must not be permitted to place or tie reinforcing steel more than 6 feet above any adjacent working surfaces unless they are protected by the use of a fall protection.

Cast-in-Place Concrete

General Requirements for Formwork   Formwork must be designed, fabricated, erected, supported, braced, and maintained so that it will be capable of supporting without failure all vertical and lateral loads that might be applied to the formwork.

Drawings or Plans   Drawings and plans, including all revisions for the jack layout, formwork (including shoring equipment), working decks, and scaffolds must be available at the jobsite.

Shoring and Reshoring   All shoring equipment (including equipment used in reshoring operations) must be inspected prior to erection to determine that the equipment meets the requirements specified in the formwork drawings.

Damaged shoring equipment must not be used for shoring. Erected shoring equipment must be inspected immediately prior to, during, and immediately after concrete placement. Shoring equipment that is found to be damaged or weakened after erection must be immediately reinforced.

If single-post shores are used one on top of another (tiered), then additional shoring requirements must be met. The shores must be as follows:

- Designed by a qualified designer and the erected shoring inspected by an engineer qualified in structural design
- Vertically aligned
- Spliced to prevent misalignment
- Adequately braced in two mutually perpendicular directions at the splice level. Each tier also must be diagonally braced in the same two directions

Adjustment of single-post shores to raise formwork must not be made after the placement of concrete.

Reshoring must be erected, as the original forms and shores are removed, whenever the concrete is required to support loads in excess of its capacity.

Vertical Slip Forms   The steel rods or pipes on which jacks climb or by which the forms are lifted must be (1) specifically designed for that purpose and (2) adequately braced where not encased in concrete. Forms must be designed to prevent excessive distortion of the structure during the jacking operation. Jacks and vertical supports must be positioned in such a manner that the loads do not exceed the rated capacity of the jacks.

The jacks or other lifting devices must be provided with mechanical dogs or other automatic holding devices to support the slip forms whenever failure of the power supply or lifting mechanisms occurs.

The form structure must be maintained within all design tolerances specified for plumpness during the jacking operation.

The predetermined safe rate of lift must not be exceeded.

All vertical slip forms must be provided with scaffolds or work platforms where employees are required to work or pass.

Reinforcing Steel   Reinforcing steel for walls, piers, columns, and similar vertical structures must be adequately supported to prevent overturning and collapse.

Employers must take measures to prevent unrolled wire mesh from recoiling. Such measures may include, but are not limited to, securing each end of the roll or turning over the roll.

Removal of Formwork   Forms and shores (except those used for slabs on grade and slip forms) must not be removed until the employer determines that the concrete has gained sufficient strength to support its weight and superimposed loads. Such determination must be based on compliance with one of the following:

- The plans and specifications stipulate conditions for removal of forms and shores and such conditions have been followed.

- The concrete has been properly tested with an appropriate American Society for Testing and Materials (ASTM) standard test method designed to indicate the concrete compressive strength, and the test results indicate that the concrete has gained sufficient strength to support its weight and superimposed loads.

Reshoring must not be removed until the concrete being supported has attained adequate strength to support its weight and all loads in place upon it.

Precast Concrete   Precast concrete wall units, structural framing, and tilt-up wall panels must be adequately supported to prevent overturning and to prevent collapse until permanent connections are completed.

Lifting inserts that are embedded or otherwise attached to tilt-up wall panels must be capable of supporting at least two times the maximum intended load applied or transmitted to them; lifting inserts for other precast members must be capable of supporting four times the load.

Only essential employees are permitted under precast concrete that is being lifted or tilted into position.

Lift-Slab Operations

- Lift-slab operations must be designed and planned by a registered professional engineer who has experience in lift-slab construction. Such plans and designs must be implemented by the employer and must include detailed instructions and sketches indicating the prescribed method of erection. The plans and designs must also include provisions for ensuring lateral stability of the building/structure during construction.

- Jacking equipment must be capable of supporting at least 2½ times the load being lifted during jacking operations and the equipment must not be overloaded. For the purpose of this provision, jacking equipment includes any load-bearing component that is used to carry out the lifting operation(s). Such equipment includes, but is not limited to, the following: threaded rods, lifting attachments, lifting nuts, hook-up collars, T-caps, spearheads, columns, and footings.

- No employee, except those essential to the jacking operation, must be permitted in the building/structure while any jacking operation is taking place unless the building/structure has been reinforced sufficiently to ensure its integrity during erection. The phrase "reinforced sufficiently to ensure its integrity" used in this paragraph means that a registered professional engineer, independent of the engineer who designed and planned the lifting operation, has determined from the plans that if there is a loss of support at any jack location, that loss will be confined to that location and the structure as a whole will remain stable.

- Under no circumstances must any employee who is not essential to the jacking operation be permitted immediately beneath a slab while it is being lifted.

Masonry Construction

Whenever a masonry wall is being constructed, employers must establish a limited access zone prior to the start of construction. The limited access zone must be as follows:

- Equal to the height of the wall to be constructed plus 4 feet and running the entire length of the wall
- On the side of the wall that will be unscaffolded
- Restricted to entry only by employees actively engaged in constructing the wall
- Kept in place until the wall is adequately supported to prevent overturning and collapse unless the height of wall is more than 8 feet and unsupported; in which case, it must be braced. The bracing must remain in place until permanent supporting elements of the structure are in place

STEEL ERECTION

Approval to Begin Steel Erection

Before authorizing the commencement of steel erection, the controlling contractor shall ensure that the steel erector is provided with the following written notifications:

- The concrete in the footings, piers, and walls and the mortar in the masonry piers and walls has attained, on the basis of an appropriate ASTM standard test method of field-cured samples, either 75% of the intended minimum compressive design strength or sufficient strength to support the loads imposed during steel erection.

- Any repairs, replacements, and modifications to the anchor bolts were conducted in accordance with the requirements listed later in this section under the heading: Repair, Replacement, or Field Modification of Anchor Rods (anchor bolts).

Commencement of Steel Erection

A steel erection contractor shall not erect steel unless it has received written notification that the concrete in the footings, piers, and walls or the mortar in the masonry piers and walls has attained, on the basis of an appropriate ASTM standard test method of field-cured samples, either 75% of the intended minimum compressive design strength or sufficient strength to support the loads imposed during steel erection.

Site Layout

The controlling contractor shall ensure that the following is provided and maintained:

- Adequate access roads into and through the site for the safe delivery and movement of derricks, cranes, trucks, other necessary equipment, and the material to be erected and means and methods for pedestrian and vehicular control. Exception: This requirement does not apply to roads outside of the construction site

- A firm, properly graded, drained area, readily accessible to the work with adequate space for the safe storage of materials and the safe operation of the erector's equipment

Preplanning of Overhead Hoisting Operations

All hoisting operations in steel erection shall be preplanned to ensure that the requirements listed later in this section under the heading Working Under Loads.

Site-Specific Erection Plan

Where employers elect, due to conditions specific to the site, to develop alternate means and methods that provide employee protection, a site-specific erection plan shall be developed by a qualified person and be available at the work site. Guidelines for establishing a site-specific erection plan are contained in Resource M, located at the back of this book and at dewalt.com/guides.

Preshift Visual Inspection of Cranes

Cranes being used in steel erection activities shall be visually inspected prior to each shift by a competent person; the inspection shall include observation for deficiencies during operation. At a minimum this inspection shall include the following:

- All control mechanisms for maladjustments

- Control and drive mechanism for excessive wear of components and contamination by lubricants, water, or other foreign matter

- Safety devices, including but not limited to boom angle indicators, boom stops, boom kickout devices, anti-two block devices, and load moment indicators where required
- Air, hydraulic, and other pressurized lines for deterioration or leakage, particularly those which flex in normal operation
- Hooks and latches for deformation, chemical damage, cracks, or wear
- Wire rope reeving for compliance with hoisting equipment manufacturer's specifications
- Electrical apparatus for malfunctioning, signs of excessive deterioration, dirt, or moisture accumulation
- Hydraulic system for proper fluid level
- Tires for proper inflation and condition
- Ground conditions around the hoisting equipment for proper support, including ground settling under and around outriggers, groundwater accumulation, or similar conditions
- The hoisting equipment for level position
- The hoisting equipment for level position after each move and setup

If any deficiency is identified, an immediate determination shall be made by the competent person as to whether the deficiency constitutes a hazard. If the deficiency is determined to constitute a hazard, the hoisting equipment shall be removed from service until the deficiency has been corrected.

The operator shall be responsible for those operations under the operator's direct control. Whenever there is any doubt as to safety, the operator shall have the authority to stop and refuse to handle loads until safety has been assured.

A qualified rigger (a rigger who is also a qualified person) shall inspect the rigging prior to each shift.

The headache ball, hook, or load shall not be used to transport personnel. Safety latches on hooks shall not be deactivated or made inoperable except:

- When a qualified rigger has determined that the hoisting and placing of purlins and single joists can be performed more safely by doing so
- When equivalent protection is provided in a site-specific erection plan.

Working Under Loads

Routes for suspended loads shall be preplanned to ensure that no employee is required to work directly below a suspended load except for:

- Employees engaged in the initial connection of the steel
- Employees necessary for the hooking or unhooking of the load

When working under suspended loads, the following criteria shall be met:

- Materials being hoisted shall be rigged to prevent unintentional displacement.
- Hooks with self-closing safety latches or their equivalent shall be used to prevent components from slipping out of the hook.
- All loads shall be rigged by a qualified rigger.

Multiple-Lift Rigging Procedure

- A multiple lift shall only be performed if the following criteria are met:
  - A multiple-lift rigging assembly is used.

- ○ A maximum of five members are hoisted per lift.
- ○ Only beams and similar structural members are lifted.
- ○ All employees engaged in the multiple lift have been trained in these procedures.
- No crane is permitted to be used for a multiple lift where such use is contrary to the manufacturer's specifications and limitations.
- Components of the multiple-lift rigging assembly shall be specifically designed and assembled with a maximum capacity for total assembly and for each individual attachment point. This capacity, certified by the manufacturer or a qualified rigger, shall be based on the manufacturer's specifications with a 5 to 1 safety factor for all components.
- The total load shall not exceed:
  - ○ The rated capacity of the hoisting equipment specified in the hoisting equipment load charts
  - ○ The rigging capacity specified in the rigging rating chart
- The multiple-lift rigging assembly shall be rigged with members:
  - ○ Attached at their center of gravity and maintained reasonably level.
  - ○ Rigged from top down
  - ○ Rigged at least 7 feet apart
- The members on the multiple-lift rigging assembly shall be set from the bottom up.
- Controlled load lowering shall be used whenever the load is over the connectors.

Structural Steel Assembly

Structural stability shall be maintained at all times during the erection process. The following additional requirements shall apply for multi-story structures:

- The permanent floors shall be installed as the erection of structural members progresses, and there shall be not more than eight stories between the erection floor and the uppermost permanent floor, except where the structural integrity is maintained as a result of the design.
- At no time shall there be more than four floors or 48 feet, whichever is less, of unfinished bolting or welding above the foundation or uppermost permanently secured floor, except where the structural integrity is maintained as a result of the design.
- A fully planked or decked floor or nets shall be maintained within two stories or 30 feet, whichever is less, directly under any erection work being performed.

Tripping Hazards

- Shear connectors (such as headed steel studs, steel bars, or steel lugs), reinforcing bars, deformed anchors, or threaded studs shall not be attached to the top flanges of beams, joists, or beam attachments so that they project vertically from or horizontally across the top flange of the member until after the metal decking, or other walking/working surface, has been installed.

Installation of Shear Connectors on Composite Floors, Roofs, and Bridge Decks

- When shear connectors are used in construction of composite floors, roofs, and bridge decks, employees shall lay out and install the shear connectors after the metal decking has been installed, using the metal decking as a working platform. Shear connectors shall not be installed from within a controlled decking zone (CDZ).

Plumbing-up

- When deemed necessary by a competent person, plumbing-up equipment shall be installed in conjunction with the steel erection process to ensure the stability of the structure.

- When used, plumbing-up equipment shall be in place and properly installed before the structure is loaded with construction material such as loads of joists, bundles of decking, or bundles of bridging.

- Plumbing-up equipment shall be removed only with the approval of a competent person.

Metal Decking

- Bundle packaging and strapping shall not be used for hoisting unless specifically designed for that purpose.

- If loose items such as dunnage, flashing, or other materials are placed on the top of metal decking bundles to be hoisted, such items shall be secured to the bundles.

- Bundles of metal decking on joists shall be landed properly.

- Metal decking bundles shall be landed on framing members so that enough support is provided to allow the bundles to be unbanded without dislodging the bundles from the supports.

- At the end of the shift or when environmental or jobsite conditions require, metal decking shall be secured against displacement.

Roof and Floor Holes and Openings

Metal decking at roof and floor holes and openings shall be installed as follows:

- Framed metal deck openings shall have structural members turned down to allow continuous deck installation except where not allowed by structural design constraints or constructability.

- Roof and floor holes and openings shall be decked over. Where large size, configuration, or other structural design does not allow openings to be decked over (such as elevator shafts, stairwells, etc.) employees shall be protected.

Metal decking holes and openings shall not be cut until immediately prior to being permanently filled with the equipment or structure needed or intended to fulfill its specific use.

Covering Roof and Floor Openings

- Covers for roof and floor openings shall be capable of supporting, without failure, twice the weight of the employees, equipment, and materials that may be imposed on the cover at any one time.

- All covers shall be secured when installed to prevent accidental displacement by the wind, equipment, or employees.

- All covers shall be painted with high-visibility paint or shall be marked with the word "HOLE" or "COVER" to provide warning of the hazard.

- Smoke dome or skylight fixtures that have been installed are not considered covers.

Decking Gaps Around Columns

- Wire mesh, exterior plywood, or equivalent shall be installed around columns where planks or metal decking do not fit tightly. The materials used must be of sufficient strength to provide fall protection for personnel and prevent objects from falling through.

Installation of Metal Decking

- Metal decking shall be laid tightly and immediately secured upon placement to prevent accidental movement or displacement. During initial placement, metal decking panels shall be placed to ensure full support by structural members.

Derrick Floors

- A derrick floor shall be fully decked and/or planked and the steel member connections completed to support the intended floor loading. Temporary loads placed on a derrick floor shall be distributed over the underlying support members so as to prevent local overloading of the deck material.

Column Anchorage

- All columns shall be anchored by a minimum of four anchor rods (anchor bolts).
- Each column anchor rod (anchor bolt) assembly, including the column-to-base plate weld and the column foundation, shall be designed to resist a minimum eccentric gravity load of 300 pounds located 18 inches from the extreme outer face of the column in each direction at the top of the column shaft.
- Columns shall be set on level finished floors, pregrouted leveling plates, leveling nuts, or shim packs that are adequate to transfer the construction loads.
- All columns shall be evaluated by a competent person to determine whether guying or bracing is needed; if guying or bracing is needed, it shall be installed.

Repair, Replacement, or Field Modification of Anchor Rods (Anchor Bolts)

- Anchor rods (anchor bolts) shall not be repaired, replaced, or field-modified without the approval of the project structural engineer of record.
- Prior to the erection of a column, the controlling contractor shall provide written notification to the steel erector if there has been any repair, replacement, or modification of the anchor rods (anchor bolts) of that column.

Beams and Columns

- During the final placing of solid web structural members, the load shall not be released from the hoisting line until the members are secured with at least two bolts per connection, of the same size and strength as shown in the erection drawings, drawn up wrench-tight or the equivalent as specified by the project structural engineer of record, unless diagonal bracing is used.
- A competent person shall determine if more than two bolts are necessary to ensure the stability of cantilevered members; if additional bolts are needed, they shall be installed.

Diagonal Bracing

- Solid web structural members used as diagonal bracing shall be secured by at least one bolt per connection drawn up wrench-tight or the equivalent as specified by the project structural engineer of record.
- Double connections at columns and/or at beam webs over a column. When two structural members on opposite sides of a column web, or a beam web over a column, are connected sharing common connection holes, at least one bolt with its wrench-tight nut shall remain connected to the first member unless a shop-attached or field-attached seat or equivalent connection device is supplied with the member to secure the first member and prevent the column from being displaced.

- If a seat or equivalent device is used, the seat (or device) shall be designed to support the load during the double connection process. It shall be adequately bolted or welded to both a supporting member and the first member before the nuts on the shared bolts are removed to make the double connection.

Column Splices

- Each column splice shall be designed to resist a minimum eccentric gravity load of 300 pounds located 18 inches from the extreme outer face of the column in each direction at the top of the column shaft.

Perimeter Columns

Perimeter columns shall not be erected unless:

- The perimeter columns extend a minimum of 48 inches above the finished floor to permit installation of perimeter safety cables prior to erection of the next tier, except where constructability does not allow.
- The perimeter columns have holes or other devices in or attached to perimeter columns at 42–45 inches above the finished floor and the midpoint between the finished floor and the top cable to permit installation of perimeter safety cables except where constructability does not allow.

Open Web Steel Joists

General   Where steel joists are used and columns are not framed in at least two directions with solid web structural steel members, a steel joist shall be field-bolted at the column to provide lateral stability to the column during erection. For the installation of this joist:

- A vertical stabilizer plate shall be provided on each column for steel joists. The plate shall be a minimum of 6-inch by 6-inch and shall extend at least 3 inches below the bottom chord of the joist with a 13/16-inch hole to provide an attachment point for guying or plumbing cables.
- The bottom chords of steel joists at columns shall be stabilized to prevent rotation during erection.
- Hoisting cables shall not be released until the seat at each end of the steel joist is field-bolted and each end of the bottom chord is restrained by the column stabilizer plate.

Where constructability does not allow a steel joist to be installed at the column:

- An alternate means of stabilizing joists shall be installed on both sides near the column and shall:
  - Provide stability
  - Be designed by a qualified person
  - Be shop installed
  - Be included in the erection drawings
- Hoisting cables shall not be released until the seat at each end of the steel joist is field-bolted and the joist is stabilized.

Where steel joists at or near columns span 60 feet or less, the joist shall be designed with sufficient strength to allow one employee to release the hoisting cable without the need for erection bridging.

Where steel joists at or near columns span more than 60 feet, the joists shall be set in tandem with all bridging installed unless an alternative method of erection, which provides equivalent stability to the steel joist, is designed by a qualified person and is included in the site-specific erection plan.

A steel joist or steel joist girder shall not be placed on any support structure unless such structure is stabilized.

When steel joist(s) are landed on a structure, they shall be secured to prevent unintentional displacement prior to installation.

No modification that affects the strength of a steel joist or steel joist girder shall be made without the approval of the project structural engineer of record.

Field-Bolted Joists   Except for steel joists that have been preassembled into panels, connections of individual steel joists to steel structures in bays of 40 feet or more shall be fabricated to allow for field-bolting during erection. These connections shall be field-bolted unless constructability does not allow.

Steel joists and steel joist girders shall not be used as anchorage points for a fall arrest system unless written approval to do so is obtained from a qualified person. A bridging terminus point shall be established before bridging is installed.

Attachment of Steel Joists and Steel Joist Girders

Each end of "K" series steel joists shall be attached to the support structure with a minimum of two $\frac{1}{8}$-inch fillet welds 1 inch long or with two $\frac{1}{2}$-inch bolts, or the equivalent.

Each end of "LH" and "DLH" series steel joists and steel joist girders shall be attached to the support structure with a minimum of two $\frac{1}{4}$-inch fillet welds 2 inches long or with two $\frac{3}{4}$-inch bolts, or the equivalent.

Each steel joist shall be attached to the support structure, at least at one end on both sides of the seat, immediately upon placement in the final erection position and before additional joists are placed.

Panels that have been preassembled from steel joists with bridging shall be attached to the structure at each corner before the hoisting cables are released.

Erection of Steel Joists

Both sides of the seat of one end of each steel joist that requires bridging shall be attached to the support structure before hoisting cables are released.

For joists over 60 feet, both ends of the joist shall be attached before the hoisting cables.

On steel joists that do not require erection bridging, only one employee shall be allowed on the joist until all bridging is installed and anchored.

Systems-Engineered Metal Buildings

Each structural column shall be anchored by a minimum of four anchor rods (anchor bolts). Rigid frames shall have 50% of their bolts or the number of bolts specified by the manufacturer (whichever is greater) installed and tightened on both sides of the web adjacent to each flange before the hoisting equipment is released.

Construction loads shall not be placed on any structural steel framework unless such framework is safely bolted, welded, or otherwise adequately secured.

In girt and eave strut-to-frame connections, when girts or eave struts share common connection holes, at least one bolt with its wrench-tight nut shall remain connected to the first member unless a manufacturer-supplied, field-attached seat or similar connection device is present to secure the first member so that the girt or eave strut is always secured against displacement.

Both ends of all steel joists or cold-formed joists shall be fully bolted and/or welded to the support structure before:

- Releasing the hoisting cables
- Allowing an employee on the joists
- Allowing any construction loads on the joists

Purlins and girts shall not be used as an anchorage point for a fall arrest system unless written approval is obtained from a qualified person. Purlins may only be used as a walking/working surface when installing safety systems, after all permanent bridging has been installed and fall protection is provided.

Construction loads may be placed only within a zone that is within 8 feet of the centerline of the primary support member.

Fall Protection During Steel Erection

General Requirements

Except for connectors and employees working in controlled decking zones, each employee engaged in a steel erection activity who is on a walking/working surface with an unprotected side or edge more than 15 feet above a lower level shall be protected from fall hazards by guardrail systems, safety net systems, personal fall arrest systems, positioning device systems, or fall restraint systems.

Perimeter Safety Cable

On multi-story structures, perimeter safety cables shall be installed at the final interior.

Connectors

Each connector shall:

- Be protected from fall hazards of more than two stories or 30 feet above a lower level, whichever is less
- Have completed specified connector training
- Be provided, at heights over 15 and up to 30 feet above a lower level, with a personal fall arrest system, positioning device system, or fall restraint system and wear the equipment necessary to be able to be tied off or be provided with other means of protection from fall hazards

Controlled Decking Zone (CDZ)

A controlled decking zone may be established in that area of the structure over 15 and up to 30 feet above a lower level where metal decking is initially being installed and forms the leading edge of a work area. In each CDZ, the following shall apply:

- Each employee working at the leading edge in a CDZ shall be protected from fall hazards of more than two stories or 30 feet, whichever is less.
- Access to a CDZ shall be limited to only those employees engaged in leading edge work.
- The boundaries of a CDZ shall be designated and clearly marked. The CDZ shall not be more than 90 feet wide and 90 feet deep from any leading edge. The CDZ shall be marked by the use of control lines or the equivalent.
- Each employee working in a CDZ shall have completed CDZ training.
- Unsecured decking in a CDZ shall not exceed 3,000 square feet.
- Safety deck attachments shall be performed in the CDZ from the leading edge back to the control line and shall have at least two attachments for each metal decking panel.
- Final deck attachments and installation of shear connectors shall not be performed in the CDZ.

Custody of Fall Protection

Fall protection provided by the steel erector shall remain in the area where steel erection activity has been completed, to be used by other trades, only if the controlling contractor or its authorized representative:

- Has directed the steel erector to leave the fall protection in place
- Has inspected and accepted control and responsibility of the fall protection prior to authorizing persons other than steel erectors to work in the area

Training

Training Personnel   Training shall be provided by a qualified person(s).

Fall Hazard Training   The employer shall train each employee exposed to a fall hazard. The employer shall institute a training program and ensure employee participation in the program. The training program shall include:

- The recognition and identification of fall hazards in the work area
- The use and operation of guardrail systems (including perimeter safety cable systems), personal fall arrest systems, positioning device systems, fall restraint systems, safety net systems, and other protection to be used
- The correct procedures for erecting, maintaining, disassembling, and inspecting the fall protection systems to be used
- The procedures to be followed to prevent falls to lower levels and through or into holes and openings in walking/working surfaces and walls
- The fall protection requirements of the steel erection standard

Special Training Programs   The employer shall provide special training to employees engaged in the following activities.

- Multiple-lift rigging procedure. The employer shall ensure that each employee who performs multiple-lift rigging has been provided training in the following areas:
  - The nature of the hazards associated with multiple lifts
  - The proper procedures and equipment to perform multiple lifts
- Connector procedures. The employer shall ensure that each connector has been provided training in the following areas:
  - The nature of the hazards associated with connecting
  - The establishment, access, proper connecting techniques, and work practices
- Controlled decking zone procedures. Where CDZs are being used, the employer shall assure that each employee has been provided training in the following areas:
  - The nature of the hazards associated with work within a controlled decking zone
  - The establishment, access, proper installation techniques, and work practices

UNDERGROUND CONSTRUCTION

The OSHA regulation for the construction of underground tunnels, shafts, chambers, and passageways can be found in Subpart S. It also applies to cut-and-cover excavations, both those physically connected to ongoing underground construction tunnels and those that create conditions characteristic of underground construction. These hazards include:

- Reduced natural ventilation and light, difficult and limited access and egress, exposure to air contaminants, fire, and explosion
- A higher accident and injury rate for workers in this industry than for other workers in the heavy construction industry

Competent Person in Underground Construction

A "competent person" is one who is capable of identifying existing and predictable hazards in the workplace and is authorized to take corrective action to eliminate them. Under this standard, a competent person is responsible for determining whether air contaminants are present in sufficient quantities to be dangerous to life; for testing the atmosphere for flammable limits before restoring power and equipment and before returning to work after a ventilation system has been shut down due to hazardous levels of flammable gas or methane; for inspecting the work area for ground stability; for inspecting all drilling equipment prior to each use; and for inspecting hauling equipment before each shift and visually checking all hoisting machinery, equipment, anchorages, and rope at the beginning of each shift and during hoisting, as necessary.

Safety Instruction

The standard requires that employees be taught to recognize and avoid hazards associated with underground construction. The instruction shall include the following topics, as appropriate for the jobsite:

- Air monitoring
- Ventilation and illumination
- Communications
- Flood control
- Mechanical and personal protective equipment
- Explosives; fire prevention and protection
- Emergency procedures—evacuation plans and check-in and check-out procedures

Access and Egress

The employer must provide safe access to and egress from all work stations and must prevent any unauthorized entry underground. Completed or unused sections of an underground work area must be barricaded. Unused openings must be covered, fenced off, or posted with warning signs indicating "Keep Out" or other, similar language.

Check-in/Check-out

The employer is required to maintain a check-in/check-out procedure that ensures that aboveground personnel can have an accurate count of the number of persons underground in an emergency. At least one designated person is to be on duty aboveground whenever anyone is working underground. This person is also responsible for securing immediate aid for and keeping an accurate count of employees underground in case of an emergency.

Hazardous Classifications

The standard provides classification criteria for gassy or potentially gassy operations and identifies additional requirements for work in gassy operations.

- **Potentially Gassy Operations:** Potentially gassy operations occur under either of the following circumstances:
  - When air monitoring shows, for more than a 24-hour period, 10% or more of the lower explosive limit (LEL) for methane or other flammable gases measured at 12 inches 1 ± 0.25 inch from the roof, face, floor, or walls in any underground work area; or

- ○ When the geological formation or history of the area shows that 10% or more of the LEL for methane or other flammable gases is likely to be encountered in the underground operation.
- **Gassy Operations:** Gassy operations occur under the following conditions:
  - ○ When air monitoring shows, for 3 consecutive days, 10% or more of the LEL for methane or other flammable gases measured at 12 inches 1 ± 0.25 inch from the roof, face, floor, or walls in any underground work area
  - ○ When methane or other flammable gases mandating from the strata have ignited, indicating the presence of such gases
  - ○ When the underground operation is connected to a currently gassy underground work area and is subject to a continuous course of air containing a flammable gas concentration

When a gassy operation exists, additional safety precautions are required. These include using more stringent ventilation requirements; using diesel equipment only if it is approved for use in gassy operations; posting each entrance with warning signs; prohibiting smoking and personal sources of ignition; maintaining a fire watch when hot work is performed; and suspending all operations in the affected area until all special requirements are met or the operation is declassified. Additional air monitoring is also required during gassy conditions.

Air Monitoring

Under the standard, the employer is required to assign a competent person to perform all air monitoring required to determine proper ventilation and quantitative measurements of potentially hazardous gases. In instances where monitoring of airborne contaminants is required by the standard to be conducted "as often as necessary," this individual is responsible for determining which substances to monitor and how frequently, taking into consideration factors such as jobsite location, geology, history, work practices, and conditions.

The atmosphere in all underground areas shall be tested quantitatively for carbon monoxide, nitrogen dioxide, hydrogen sulfide, and other toxic gases, dusts, vapors, mists, and fumes as often as necessary to ensure that prescribed limits are met. Quantitative tests for methane shall also be performed in order to determine whether an operation is gassy or potentially gassy.

A record of all air-quality tests (including location, date, time, substances, and amount monitored) is to be kept aboveground at the worksite and shall be made available to the Secretary of Labor upon request.

Oxygen

Testing is to be performed as often as necessary to assure that the atmosphere at normal atmospheric pressure contains at least 19.5% oxygen, but not more than 22.5%.

Hydrogen Sulfide

When air monitoring indicates the presence of 5 parts per million (ppm) or more of hydrogen sulfide, testing is to be conducted in the affected area at the beginning and midpoint of each shift until the concentration of hydrogen sulfide has been less than 5 ppm for 3 consecutive days. Continuous monitoring shall be performed when hydrogen sulfide is present above 10 ppm. Employees must be notified when the concentration of hydrogen sulfide is above 10 ppm. At concentrations of 20 ppm, an alarm (visual and aural) must signal to indicate that additional measures might be required (e.g., respirators, increased ventilation, evacuation) to maintain the proper exposure levels.

Other Precautions

When the competent person determines that there are contaminants present that are dangerous to life, the employer must post notices of the condition at all entrances to underground work areas and must ensure that the necessary precautions are taken.

In cases where 5% or more of the LEL for these gases is present, steps must be taken to increase ventilation air volume to reduce the concentration to less than 5% of the LEL (except when operating under gassy/potentially gassy requirements).

When 10% or more of the LEL for methane or other flammable gases is detected where welding, cutting, or other "hot" work is being performed, work shall be suspended until the concentration is reduced to less than 10% of the LEL.

Where there is a concentration of 20% or more LEL, all employees shall be immediately withdrawn to a safe location aboveground, except those necessary to eliminate the hazard, and electrical power, except for acceptable pumping and ventilating equipment, shall be cut off to the endangered area until the concentration of the gas is less than 20% of the LEL.

Potentially gassy and gassy operations require additional air monitoring. These include testing for oxygen in the affected work areas; using flammable-gas monitoring equipment (continuous automatic when using rapid excavation machines; manual as needed to monitor prescribed limits); performing local gas tests prior to doing, and continuously during, any hot work; testing continuously for flammable gas when employees are working underground using drill and blast methods and prior to reentry after blasting.

Ventilation

There are a number of requirements for ventilation in underground construction activities. In general, fresh air must be supplied to all underground work areas in sufficient amounts to prevent any dangerous or harmful accumulation of dusts, fumes, mists, vapors, or gases. A minimum of 200 cubic feet of fresh air per minute is to be supplied for each employee underground. Mechanical ventilation, with reversible airflow, is to be provided in all of these work areas, except where natural ventilation is demonstrably sufficient. Where blasting or drilling is performed or other types of work operations that may cause harmful amounts of dust, fumes, vapors, etc., the velocity of airflow must be at least 30 feet per minute.

For gassy or potentially gassy operations, ventilation systems must meet additional requirements. Ventilation systems used during gassy operations also must have controls located aboveground for reversing airflow.

Illumination

As in all construction operations, the standard requires that proper illumination be provided during tunneling operations. When explosives are handled, only acceptable portable lighting equipment shall be used within 50 feet of any underground heading.

Fire Prevention and Control

Open flames and fires are prohibited in all underground construction activities, except for hot work operations. Smoking is allowed only in areas free of fire and explosion hazards, and the employer is required to post signs prohibiting smoking and open flames where these hazards exist. Various work practices are also identified as preventive measures. For example, there are limitations on the piping of diesel fuel from the surface to an underground location. Also, the pipe or hose system used to transfer fuel from the surface to the storage tank must remain empty except when transferring the fuel. Gasoline is not to be used, stored, or carried underground. Gases such as acetylene, liquefied petroleum, and methylacetylenepropadiene (stabilized) may be used underground only for hot work operations. Leaks and spills of flammable or combustible fluids must be cleaned up immediately. The standard also requires fire prevention measures regarding fire-resistant barriers, fire-resistant hydraulic fluids, the location and storage of combustible materials near openings or access to underground operations, electrical installations underground, lighting fixtures, fire extinguishers, etc.

Hot Work

During hot work operations such as welding, noncombustible barriers must be installed below work being performed in or over a shaft or raise. As mentioned earlier, during these operations, only the amount of fuel gas and oxygen cylinders necessary to perform welding, cutting, or other hot work over the next 24-hour period shall be kept underground. When work is completed, gas and oxygen cylinders shall be removed.

Emergencies

At worksites where 25 or more employees work underground at one time, employers are required to provide rescue teams or rescue services, which include at least two 5-person teams (one on the jobsite or within ½-hour travel time and one within 2 hours travel time). Where there are fewer than 25 employees underground at one time, the employer shall provide or make available in advance one 5-person rescue time on site or within ½-hour travel time.

Rescue team members have to be qualified in rescue procedures and in the use of firefighting equipment and breathing apparatus. Their qualifications must be reviewed annually.

The employer must ensure that rescue teams are familiar with the jobsite conditions. Rescue team members are required to practice donning and using self-contained breathing apparatus on a monthly basis for jobsites where flammable or noxious gases are encountered or anticipated in hazardous quantities.

As part of emergency procedures, the employer shall provide self-rescuers (currently approved by NIOSH and MSHA) to be immediately available to all employees at underground work stations who might be trapped by smoke or gas.

A "designated," or authorized, person shall be responsible for securing immediate aid for workers and for keeping an accurate count of employees underground. Emergency lighting, a portable hand or cap lamp, shall be provided to all underground workers in their work areas to provide adequate light for escape.

DEMOLITION

Before the start of every demolition job, the demolition contractor should take a number of steps to safeguard the health and safety of workers at the jobsite. These preparatory operations involve the overall planning of the demolition job, including the methods to be used to bring the structure down, the equipment necessary to do the job, and the measures to be taken to perform the work safely. Planning for a demolition job is as important as actually doing the work. Therefore, all planning work should be performed by a competent person experienced in all phases of the demolition work to be performed.

The American National Standards Institute (ANSI) in its ANSI A10.6-1983–Safety Requirements For Demolition Operations states:

"No employee shall be permitted in any area that can be adversely affected when demolition operations are being performed. Only those employees necessary for the performance of the operations shall be permitted in these areas."

Engineering Survey

Prior to starting all demolition operations, OSHA Standard 1926.850(a) requires that an engineering survey of the structure must be conducted by a competent person. The purpose of this survey is to determine the condition of the framing, floors, and walls so that measures can be taken, if necessary, to prevent the premature collapse of any portion of the structure. When indicated as advisable, any adjacent structure(s) or improvements should also be similarly checked. The demolition contractor must maintain a written copy of this survey. Photographing existing damage in neighboring structures is also advisable. The engineering survey provides the demolition contractor with the opportunity to evaluate the job in its entirety. The contractor should plan for the wrecking of the structure, the equipment to do the work, manpower requirements, and the protection of the public. The safety of all workers on the jobsite should be a prime consideration. During the preparation of the engineering survey, the contractor should plan for potential hazards such as fires, cave-ins, and injuries.

If the structure to be demolished has been damaged by fire, flood, explosion, or some other cause, appropriate measures, including bracing and shoring of walls and floors, shall be taken to protect workers and any adjacent structures. It shall also be determined if any type of hazardous chemicals, gases, explosives, flammable material, or similar dangerous substances have been used or stored on the site. If the nature of a substance cannot be easily determined, samples should be taken and analyzed by a qualified person prior to demolition.

During the planning stage of the job, all safety equipment needs should be determined. The required number and type of respirators, lifelines, warning signs, safety nets, special face and eye protection, hearing protection, and other worker protection devices should be determined during the preparation of the engineering survey. A comprehensive plan is necessary for any confined-space entry.

Utility Location

One of the most important elements of the prejob planning is the location of all utility services. All electric, gas, water, steam, sewer, and other service lines should be shut off, capped, or otherwise controlled at or outside the building before demolition work is started. In each case, any utility company involved should be notified in advance, and its approval or services, if necessary, shall be obtained.

If it is necessary to maintain any power, water, or other utilities during demolition, such lines shall be temporarily relocated as necessary and/or protected. The location of all overhead power sources should also be determined, as they can prove especially hazardous during any machine demolition. All workers should be informed of the location of any existing or relocated utility service.

Medical Services and First Aid

Prior to starting work, provisions should be made for prompt medical attention in case of serious injury. The nearest hospital, infirmary, clinic, or physician shall be located as part of the engineering survey. The job supervisor should be provided with instructions for the most direct route to these facilities. Proper equipment for prompt transportation of an injured worker, as well as a communication system to contact any necessary ambulance service, must be available at the jobsite. The telephone numbers of the hospitals, physicians, or ambulances shall be conspicuously posted.

In the absence of an infirmary, clinic, hospital, or physician that is reasonably accessible in terms of time and distance to the worksite, a person who has a valid certificate in first-aid training from the U.S. Bureau of Mines, the American Red Cross, or equivalent training should be available at the worksite to render first aid.

A properly stocked first-aid kit as determined by an occupational physician must be available at the jobsite. The first-aid kit should contain approved supplies in a weatherproof container with individual sealed packages for each type of item. It should also include rubber gloves to prevent the transfer of infectious diseases.

Provisions should also be made to provide for quick drenching or flushing of the eyes should any person be working around corrosive materials. Eye flushing must be done with water containing no additives. The contents of the kit shall be checked before being sent out on each job and at least weekly to ensure the expended items are replaced.

Police and Fire Contact

The telephone numbers of the local police, ambulance, and fire departments should be available at each jobsite. This information can prove useful to the job supervisor in the event of any traffic problems, such as the movement of equipment to the job, uncontrolled fires, or other police/fire matters. The police number may also be used to report any vandalism, unlawful entry to the jobsite, or accidents requiring police assistance.

Fire Prevention and Protection

A "fire plan" should be set up prior to beginning a demolition job. This plan should outline the assignments of key personnel in the event of a fire and provide an evacuation plan for workers on the site.

Common sense should be the general rule in all fire prevention planning:

- All potential sources of ignition should be evaluated and the necessary corrective measures taken.

- Electrical wiring and equipment for providing light, heat, or power should be installed by a competent person and inspected regularly.

- Equipment powered by an internal combustion engine should be located so that the exhausts discharge well away from combustible materials and away from workers.

- When the exhausts are piped outside the building, a clearance of at least 6 inches should be maintained between such piping and combustible material.

- All internal combustion equipment should be shut down prior to refueling. Fuel for this equipment should be stored in a safe location.

- Sufficient firefighting equipment should be located near any flammable or combustible liquid storage area.

- Only approved containers and portable tanks should be used for the storage and handling of flammable and combustible liquids.

- Heating devices should be situated so they are not likely to overturn and shall be installed in accordance with their listing, including clearance to combustible material or equipment. Temporary heating equipment, when utilized, should be maintained by competent personnel.

- Smoking should be prohibited at or in the vicinity of hazardous operations or materials. Where smoking is permitted, safe receptacles shall be provided for smoking materials.

- Roadways between and around combustible storage piles should be at least 15 feet wide and maintained free from accumulation of rubbish, equipment, or other materials.

- When storing debris or combustible material inside a structure, such storage shall not obstruct or adversely affect the means of exit.

- A suitable location at the jobsite should be designated and provided with plans, emergency information, and equipment, as needed. Access for heavy firefighting equipment should be provided on the immediate jobsite at the start of the job and maintained until the job is completed.

- Free access from the street to fire hydrants and to outside connections for standpipes, sprinklers, or other fire extinguishing equipment, whether permanent or temporary, should be provided and maintained at all times.

- Pedestrian walkways should not be so constructed as to impede access to hydrants.

- No material or construction should interfere with access to hydrants, siamese connections, or fire extinguishing equipment.

- A temporary or permanent water supply of sufficient volume, duration, and pressure, required to properly operate the firefighting equipment, should be made available.

- Standpipes with outlets should be provided on large multi-story buildings to provide for fire protection on upper levels. If the water pressure is insufficient, a pump should also be provided.

- An ample number of fully charged portable fire extinguishers should be provided throughout the operation. All motor-driven mobile equipment should be equipped with an approved fire extinguisher.

- An alarm system (e.g., telephone system, siren, two-way radio, etc.) shall be established in such a way that employees on the site and the local fire department can be alerted in case of an emergency. The alarm code and reporting instructions shall be conspicuously posted and the alarm system should be serviceable at the jobsite during the demolition. Fire cut-offs shall be retained in the buildings undergoing alterations or demolition until operations necessitate their removal.

Demolishing a Chimney, Stack, Silo, or Cooling Tower

Inspection and Planning

When preparing to demolish any chimney, stack, silo, or cooling tower, the first step must be a careful, detailed inspection of the structure by an experienced person. If possible, architectural/engineering drawings should be consulted. Particular attention should be paid to the condition of the chimney or stack. Workers should be on the lookout for any structural defects such as weak or acid-laden mortar joints and any cracks or openings. The interior brickwork in some sections of industrial chimney shafts can be extremely weak. If the stack has been banded with steel straps, these bands shall be removed only as the work progresses from the top down. Sectioning of the chimney by water, etc., should be considered.

Safe Work Practice

When hand demolition is required, it should be carried out from a working platform.

Experienced personnel must install a self-supporting tubular scaffold, suspended platform, or knee-braced scaffolding around the chimney. Particular attention should be paid to the design, support, and tie-in (braces) of the scaffold.

A competent person should be present at all times during the erection of the scaffold.

It is essential that there be adequate working clearance between the chimney and the work platform.

Access to the top of the scaffold should be provided by means of portable walkways.

The platforms should be decked solid and the area from the work platform to wall bridged with a minimum of 2-inch-thick lumber.

A back rail 42 inches above the platform with a midrail covered with canvas or mesh should be installed around the perimeter of the platform to prevent injury to workers below. Debris netting may be installed below the work platform.

Excess canvas or plywood attachments can form a wind sail that could cause collapse of the scaffold.

When working on the work platform, all personnel should wear hard hats; long-sleeve shirts; eye/face protection, such as goggles and face shields; respirators; and safety belts, as required.

Care should be taken that the proper number of workers are assigned to the task.

Too many people on a small work platform can lead to accidents.

An alternative to the erection of a self-supporting tubular steel scaffold is to "climb" the structure with a creeping bracket scaffold. Careful inspection of the masonry and a decision as to the safety of this alternative must be made by a competent person. It is essential that the masonry of the chimney be in good enough condition to support the bracket scaffold.

The area around the chimney should be roped off or barricaded and secured, with appropriate warning signs posted. No unauthorized entry should be permitted to this area. It's also good practice to keep a worker (i.e., a supervisor, operating engineer, another worker, or a "safety person") on the ground with a means of communication to the workers above.

Special attention should be paid to weather conditions when working on a chimney. No work should be done during inclement weather such as during lightning or high wind situations. The worksite should be wetted down, as needed, to control dust.

Debris Clearance

If debris is dropped inside the shaft, it can be removed through an opening in the chimney at grade level. The opening at grade must be kept relatively small in order not to weaken the structure. If a larger opening is desired, a professional engineer should be consulted.

When removing debris by hand, an overhead canopy of adequate strength should be provided. If machines are used for removal of debris, proper overhead protection for the operator should be used.

Excessive debris should not be allowed to accumulate inside or outside the shaft of the chimney as the excess weight of the debris can impose pressure on the wall of the structure and might cause the shaft to collapse.

The foreman should determine when debris is to be removed, halt all demolition during debris removal, and make sure the area is clear of clean-up workers before continuing demolition.

Demolition by Deliberate Collapse

Another method of demolishing a chimney or stack is by deliberate collapse. Deliberate collapse requires extensive planning and experienced personnel and should be used only when conditions are favorable.

There must be a clear space for the fall of the structure of at least 45 degrees on each side of the intended fall line and 1½ times the total height of the chimney. Considerable vibration may be set up when the chimney falls, so there should be no sewers or underground services on the line of the fall. Lookouts must be posted on the site and warning signals must be arranged. The public and other workers at the jobsite must be kept well back from the fall area.

The use of explosives is one way of setting off deliberate collapse. This type of demolition should be undertaken only by qualified persons. The entire work area shall be cleared of nonessential personnel before any explosives are placed. Though the use of explosives is a convenient method of bringing down a chimney or stack, there is a considerable amount of vibration produced, and caution should be taken if there is any likelihood of damage.

Demolition of Pre-Stressed Concrete Structures

The different forms of construction used in a number of more or less conventional structures built during the last few decades will give rise to a variety of problems when the time comes for them to be demolished. Pre-stressed concrete structures fall in this general category. The most important aspect of demolishing a pre-stressed concrete structure takes place during the engineering survey. During the survey, a qualified person should determine if the structure to be demolished contains any pre-stressed members. It is the responsibility of the demolition contractor to inform all workers on the demolition job site of the presence of pre-stressed concrete members within the structure. They should also instruct them in the safe work practice that must be followed to safely perform the demolition. Workers should be informed of the hazards of deviating from the prescribed procedures and the importance of following their supervisor's instruction.

Categories of Pre-Stressed Construction   There are four main categories of pre-stressed members. The category, or categories, should be determined before attempting demolition, bearing in mind that any pre-stressed structure may contain elements of more than one category.

- **Category 1.** Members pre-stressed before the application of the superimposed loads and having all cables or tendons fully bonded to the concrete or grouted within ducts.

- **Category 2.** Like Category 1, but having the tendons left ungrouted. This type of construction can sometimes be recognized from the access points that may have been provided for inspection of the cables and anchors. More recently, unbonded tendons have been used in the construction of beams, slabs, and other members; these tendons are protected by grease and surrounded by plastic sheathing instead of the usual metal duct.

- **Category 3.** Members that are pre-stressed progressively as the building construction proceeds and the dead load increases, using bonded tendons, as Category 1.

- **Category 4.** Like Category 3, but using unbonded tendons, like Category 2.

Examples of construction using members of Categories 3 or 4 are relatively rare up to this time. However, they may be found, for example, in the podium of a tall building or some types of bridges. They require that particular care be taken in demolition.

Pre-Tensioned Members   These usually do not have any end anchors, the wires being embedded or bonded within the length of the member. Simple pre-tensioned beams and slabs of spans up to about 23 feet can be demolished in a manner similar to ordinary reinforced concrete. Pre-tensioned beams and slabs may be lifted and lowered to the ground as complete units after the removal of any composite concrete covering to tops and ends of the units. To facilitate breaking up, the members should be turned on their sides. Lifting from the structure should generally be done from points near the ends of the units or from lifting point positions. Reuse of lifting eyes, if in good condition, is recommended whenever possible. When units are too large to be removed, consideration should be given to temporary supporting arrangements.

Pre-Cast Units Stressed Separately from the Main Frames of the Structure, with End Anchors and Grouted and Ungrouted Ducts   Before breaking up, units of this type should be lowered to the ground, if possible. It is advisable to seek the counsel of a professional engineer before carrying out this work, especially where there are ungrouted tendons. In general, this is true because grouting is not always 100% efficient. After lowering, the units can be turned on their side with the ends up on blocks after any composite concrete is removed. This may suffice to break the unit and release the pre-stress; if not, a sandbag screen, timbers, or a blast mat as a screen should be erected around the ends and demolition commenced, taking care to clear the area of any personnel. It should be borne in mind that the end blocks may be heavily reinforced and difficult to break up.

Monolithic Structures   The advice of the professional engineer experienced in pre-stressed work should be sought before any attempt is made to expose the tendons or anchorages of structures in which two or more members have been stressed together. It will usually be necessary for temporary supports to be provided so the tendons and the anchorage can be cautiously exposed. In these circumstances it is essential that indiscriminate attempts to expose and destress the tendons and anchorages are not made.

Progressively Pre-Stressed Structures   In the case of progressively pre-stressed structures, it is essential to obtain the advice of a professional engineer and to demolish the structure in strict accordance with the engineer's method of demolition. The stored energy in this type of structure is large. In some cases, the inherent properties of the stressed section may delay failure for some time, but the presence of these large pre-stressing forces may cause sudden and complete collapse with little warning.

Safe Work Practices When Working in Confined Spaces

Demolition contractors often come in contact with confined spaces when demolishing structures at industrial sites. These confined spaces can be generally categorized in two major groups: those with open tops and a depth that restricts the natural movement of air and enclosed spaces with very limited openings for entry. Examples of these spaces include storage tanks, vessels, degreasers, pits vaults, casing, and silos.

The hazards encountered when entering and working in confined spaces are capable of causing bodily injury, illness, and death. Accidents occur among workers because of failure to recognize that a confined space is a potential hazard. It should therefore be considered that the most unfavorable situation exists in every case and that the danger of explosion, poisoning, and asphyxiation will be present at the onset of entry.

Safe Blasting Procedures

Prior to the blasting of any structure or portion thereof, a complete written survey must be made by a qualified person of all adjacent improvements and underground utilities. When there is a possibility of excessive vibration due to blasting operations, seismic or vibration tests should be taken to determine proper safety limits to prevent damage to adjacent or nearby buildings, utilities, or other property.

The preparation of a structure for demolition by explosives may require the removal of structural columns, beams, or other building components. This work should be directed by a structural engineer or a competent person qualified to direct the removal of these structural elements. Extreme caution must be taken during this preparatory work to prevent the weakening and premature collapse of the structure.

The use of explosives to demolish smokestacks, silos, cooling towers, or similar structures should only be permitted if there is a minimum of 90 degrees of open space extending for at least 150% of the height of the structure or if the explosives specialist can demonstrate consistent previous performance with tighter constraints at the site.

Fire Precautions

The presence of fire near explosives presents a severe danger. Every effort should be made to ensure that fires or sparks do not occur near explosive materials. Smoking, matches, firearms, open-flame lamps, and other fires, flame, or heat-producing devices must be prohibited in or near explosive magazines or in areas where explosives are being handled, transported, or used. In fact, persons working near explosives should not even carry matches, lighters, or other sources of sparks or flame. Open fires or flames should be prohibited within 100 feet of any explosive materials. In the event of a fire that is in imminent danger of contact with explosives, all employees must be removed to a safe area.

Electrical detonators can be inadvertently triggered by stray RF (radio frequency) signals from two-way radios. RF signal sources should be restricted from or near to the demolition site if electrical detonators are used.

Personnel Selection

A blaster is a competent person who uses explosives. A blaster must be qualified by reason of training, knowledge, or experience in the field of transporting, storing, handling, and using explosives. In addition, the blaster should have a working knowledge of state and local regulations that pertain to explosives. Training courses are often available from manufacturers of explosives and blasting safety manuals are offered by the Institute of Makers of Explosives (IME) as well as other organizations.

Blasters shall be required to furnish satisfactory evidence of competency in handling explosives and in safely performing the type of blasting required. A competent person should always be in charge of explosives and should be held responsible for enforcing all recommended safety precautions in connection with them.

Transportation of Explosives

Vehicles used for transporting explosives shall be strong enough to carry the load without difficulty and shall be in good mechanical condition. All vehicles used for the transportation of explosives shall have tight floors, and any exposed spark-producing metal on the inside of the body shall be covered with wood or some other nonsparking material. Vehicles or conveyances transporting explosives shall only be driven by, and shall be under the supervision of, a licensed driver familiar with the local, state, and federal regulations governing the transportation of explosives. No passengers should be allowed in any vehicle transporting explosives.

Explosives, blasting agents, and blasting supplies shall not be transported with other materials or cargoes. Blasting caps shall not be transported with other materials or cargoes. Blasting caps shall not be transported in the same vehicle with other explosives. If an open-bodied truck is used, the entire load should be completely covered with a fire- and water-resistant tarpaulin to protect it from the elements. Vehicles carrying explosives should not be loaded beyond the manufacturer's safe capacity rating, and in no case should the explosives be piled higher than the closed sides and ends of the body.

Every motor vehicle or conveyance used for transporting explosives shall be marked or placarded with warning signs required by OSHA and the DOT.

Each vehicle used for transportation of explosives shall be equipped minimally with at least a 10-pound rated, serviceable, ABC fire extinguisher. All drivers should be trained in the use of the extinguishers on their vehicle.

In transporting explosives, congested traffic and high-density population areas should be avoided, where possible, and no unnecessary stops should be made. Vehicles carrying explosives, blasting agents, or blasting supplies shall not be taken inside a garage or shop for repairs or servicing. No motor vehicle transporting explosives shall be left unattended.

Storage of Explosives

All explosives must be accounted for at all times and all not being used must be kept in a locked magazine. A complete detailed inventory of all explosives received and placed in, removed from, and returned to the magazine should be maintained at all times. Appropriate authorities must be notified of any loss, theft, or unauthorized entry into a magazine.

Manufacturers' instructions for the safe handling and storage of explosives are ordinarily enclosed in each case of explosives. It is best to refer to these instructions and the aforementioned IME manuals for the specifics of storage and handling. They should be carefully followed. Packages of explosives should not be handled roughly. Sparking metal tools should not be used to open wooden cases. Metallic slitters may be used for opening fiberboard cases, provided the metallic slitter does not come in contact with the metallic fasteners of the case.

The oldest stock should always be used first to minimize the chance of deterioration from long storage. Loose explosives or broken, defective, or leaking packages can be hazardous and should be segregated and properly disposed of in accordance with the specific instructions of the manufacturer. If the explosives are in good condition, it may be advisable to repack them. In this case, the explosives supplier should be contacted. Explosives cases should not be opened or explosives packed or repacked while in a magazine.

Storage Conditions

Providing a dry, well-ventilated place for the storage of explosives is one of the most important and effective safety measures. Exposure to weather damages most kinds of explosives, especially dynamite and caps. Every precaution should be taken to keep them dry and relatively cool. Dampness or excess humidity may be the cause of misfires resulting in injury or loss of life. Explosives should be stored in properly constructed fire- and bullet-resistant structures, located according to the IME American Table of Distances and kept locked at all times except when opened for use by an authorized person.

Explosives should not be left, kept, or stored where children, unauthorized persons, or animals have access to them, nor should they be stored in or near a residence.

Detonators should be stored in a separate magazine located according to the IME American Table of Distances. **Detonators should never be stored in the same magazine with any other kind of explosives.**

Ideally, arrangements should be made whereby the supplier delivers the explosives to the jobsite in quantities that will be used up during the workday. An alternative would be for the supplier to return to pick up unused quantities of explosives. If it is necessary for the contractor to store his explosives, he should be familiar with all local requirements for such storage.

Proper Use of Explosives

Blasting operations shall be conducted between sunup and sundown, whenever possible. Adequate signs should be posted to raise the alert to the hazard presented by blasting. Blasting mats or other containment should be used where there is danger of rocks or other debris being thrown into the air or where there are buildings or transportation systems nearby. Care should be taken to make sure mats and other protection do not disturb the connections to electrical blasting caps.

Radio, television, and radar transmitters create fields of electrical energy that can, under exceptional circumstances, detonate electric blasting caps. Certain precautions must be taken to prevent accidental discharge of electric blasting caps from current induced by radar, radio transmitters, lightning, adjacent power lines, dust storms, or other sources of extraneous or static electricity. These precautions shall include:

- Ensuring that mobile radio transmitters on the jobsite that are less than 100 feet away from electric blasting caps, in other than original containers, shall be de-energized and effectively locked
- Prominent display of adequate signs, warning against the use of mobile radio transmitters, on all roads within 1,000 feet of the blasting operations
- Maintaining the minimum distances recommended by the IMES between the nearest transmitter and electric blasting caps
- Suspending all blasting operations and removing persons from the blasting area during the approach and progress of an electric storm

After loading is completed, there should be as little delay as possible before firing. Each blast should be fired under the direct supervision of the blaster, who should inspect all connections before firing and who should personally see that all persons are in the clear before giving the order to fire and that standard signals, which indicate that a blast is about to be fired and a later all-clear signal, have been adopted. It is important that everyone working in the area be familiar with these signals and that they be strictly obeyed.

Procedures After Blasting

Immediately after the blast has been fired, the firing line shall be disconnected from the blasting machine and short-circuited. Where power switches are used, they shall be locked open or in the off position. Sufficient time shall be allowed for dust, smoke, and fumes to leave the blasted area before returning the spot. An inspection of the area and the surrounding rubble shall be made by the blaster to determine if all charges have been exploded before employees are allowed to return to the operation. Ail wires should be traced and the search for unexploded cartridges made by the blaster.

Disposal of Explosives

Explosives, blasting agents, and blasting supplies that are obviously deteriorated or damaged should not be used; they should be properly disposed of. Explosives distributors will usually take back old stock. Local fire marshals or representatives of the United States Bureau of Mines may also arrange for its disposal. Under no circumstances should any explosives be abandoned.

Wood, paper, fiber, or other materials that have previously contained high explosives should not be used again for any purpose, but should be destroyed by burning. These materials should not be burned in a stove, fireplace, or other confined space. Rather, they should be burned at an isolated outdoor location, at a safe distance from thoroughfares, magazines, and other structures. It is important to check that the containers are entirely empty before burning. During burning, the area should be adequately protected from intruders and all persons kept at least 100 feet from the fire.

STAIRWAYS AND LADDERS

Stairways and ladders are a major source of injuries and fatalities among construction workers. OSHA estimates that there are 24,882 injuries and as many as 36 fatalities per year due to falls from stairways and ladders used in construction. Nearly half of these injuries are serious enough to require time off the job—11,570 lost workday injuries and 13,312 non-lost-workday injuries occur annually due to falls from stairways and ladders used in construction. These data demonstrate that work on and around ladders and stairways is hazardous. More importantly, they show that compliance with OSHA's requirements for the safe use of ladders and stairways could have prevented many of these injuries.

The OSHA rules apply to all stairways and ladders used in construction, alteration, repair (including painting and decorating), and demolition worksites covered by OSHA's construction safety and health standards. They also specify when stairways and ladders must be provided. They do not apply to ladders that are specifically manufactured for scaffold access and egress, but do apply to job-made and manufactured portable ladders intended for general-purpose use and that are then used for scaffold access and egress.

General Requirements

- A stairway or ladder must be provided at all worker points of access where there is a break in elevation of 19 inches or more and no ramp, runway, embankment, or personnel hoist is provided.

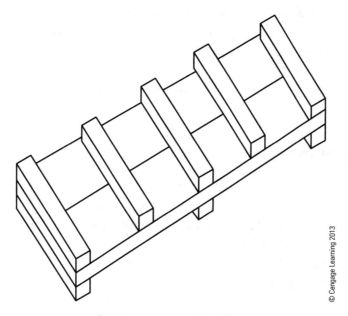

© Cengage Learning 2013

Proper ramp construction

- When there is only one point of access between levels, it must be kept clear to permit free passage by workers. If free passage becomes restricted, a second point of access must be provided and used.
- When there are more than two points of access between levels, at least one point of access must be kept clear.

- All stairway and ladder fall protection systems required by these rules must be installed and all duties required by the stairway and ladder rules must be performed before employees begin work that requires them to use stairways or ladders and their respective fall protection systems.

Stairways

The following general requirements apply to all stairways used during the process of construction, as indicated:

- Stairways that will not be a permanent part of the structure on which construction work is performed must have landings at least 30 inches deep and 22 inches wide at every 12 feet or less of vertical rise.

- Stairways must be installed at least 30 degrees, and no more than 50 degrees, from the horizontal.

- Variations in riser height or stair tread depth must not exceed ¼ inch in any stairway system, including any foundation structure used as one or more treads of the stairs.

- Where doors or gates open directly onto a stairway, a platform must be provided that is at least 20 inches in width beyond the swing of the door.

- Metal pan landings and metal pan treads must be secured in place before filling.

- All stairway parts must be free of dangerous projections such as protruding nails.

- Slippery conditions on stairways must be corrected.

- Spiral stairways that will not be a permanent part of the structure may not be used by workers.

The following requirements apply to stairs in temporary service during construction:

- Except during construction of the actual stairway, stairways with metal pan landings and treads must not be used where the treads and/or landings have not been filled in with concrete or other material, unless the pans of the stairs and/or landings are temporarily filled in with wood or other material. All treads and landings must be replaced when worn below the top edge of the pan.

- Except during construction of the actual stairway, skeleton metal frame structures and steps must not be used (where treads and/or landings are to be installed at a later date) unless the stairs are fitted with secured temporary treads and landings.

- Temporary treads must be made of wood or other solid material and installed the full width and depth of the stair.

Stair Rails and Handrails

The following general requirements apply to all stair rails and handrails:

- Stairways having four or more risers, or rising more than 30 inches in height, whichever is less, must have at least one handrail. A stair rail also must be installed along each unprotected side or edge. When the top edge of a stair rail system also serves as a handrail, the height of the top edge must not be more than 37 inches nor less than 36 inches from the upper surface of the stair rail to the surface of the tread.

- Winding or spiral stairways must be equipped with a handrail to prevent using areas where the tread width is less than 6 inches.

- Stair rails installed after March 15, 1991, must not be less than 36 inches in height.

- Midrails, screens, mesh, intermediate vertical members, or equivalent intermediate structural members must be provided between the top rail and stairway steps of the stair rail system.

- Midrails, when used, must be located midway between the top of the stair rail system and the stairway steps.

- Screens or mesh, when used, must extend from the top rail to the stairway step and along the opening between top rail supports.

- Intermediate vertical members, such as balusters, when used, must not be more than 19 inches apart.

- Other intermediate structural members, when used, must be installed so that there are no openings of more than 19 inches wide.

- Handrails and the top rails of the stair rail systems must be capable of withstanding, without failure, at least 200 pounds of weight applied within 2 inches of the top edge in any downward or outward direction, at any point along the top edge.

- The height of handrails must not be more than 37 inches nor less than 30 inches from the upper surface of the handrail to the surface of the tread.

- The height of the top edge of a stair rail system used as a handrail must not be more than 37 inches nor less than 36 inches from the upper surface of the stair rail system to the surface of the tread.

- Stair rail systems and handrails must be surfaced to prevent injuries such as punctures or lacerations and to keep clothing from snagging.

- Handrails must provide an adequate handhold for employees to grasp to prevent falls.

- The ends of stair rail systems and handrails must be constructed to prevent dangerous projections such as rails protruding beyond the end posts of the system.

- Temporary handrails must have a minimum clearance of 3 inches between the handrail and walls, stair rail systems, and other objects.

- Unprotected sides and edges of stairway landings must be provided with standard 42-inch guardrail systems.

Ladders

The following general requirements apply to all ladders, including job-made ladders:

- A double-cleated ladder or two or more ladders must be provided when ladders are the only way to enter or exit a work area having 25 or more employees or when a ladder serves simultaneous two-way traffic.

- Ladder rungs, cleats, and steps must be parallel, level, and uniformly spaced when the ladder is in position for use.

- Rungs, cleats, and steps of portable and fixed ladders (except as provided below) must not be spaced less than 10 inches apart, nor more than 14 inches apart, along the ladder's side rails.

- Rungs, cleats, and steps of step stools must not be less than 8 inches apart, nor more than 12 inches apart, between centerlines of the rungs, cleats, and steps.

- Rungs, cleats, and steps at the base section of extension trestle ladders must not be less than 8 inches nor more than 18 inches apart, between centerlines of the rungs, cleats, and steps. The rung spacing on the extension section must not be less than 6 inches nor more than 12 inches.

- Ladders must not be tied or fastened together to create longer sections unless they are specifically designed for such use.

- A metal spreader or locking device must be provided on each stepladder to hold the front and back sections in an open position when the ladder is being used.

- When splicing side rails, the resulting side rail must be equivalent in strength to a one-piece side rail made of the same material.

- Two or more separate ladders used to reach an elevated work area must be offset with a platform or landing between the ladders, except when portable ladders are used to gain access to fixed ladders.

- Ladder components must be surfaced to prevent injury from punctures or lacerations and prevent snagging of clothing.
- Wood ladders must not be coated with any opaque covering, except for identification or warning labels which may be placed only on one face of a side rail.

Portable Ladders

- Non-self-supporting and self-supporting portable ladders must support at least four times the maximum intended load; extra-heavy-duty type 1A metal or plastic ladders must sustain 3.3 times the maximum intended load. The ability of a self-supporting ladder to sustain loads must be determined by applying the load to the ladder in a downward vertical direction. The ability of a non-self-supporting ladder to sustain loads must be determined by applying the load in a downward vertical direction when the ladder is placed at a horizontal angle of 75.5 degrees.
- The minimum clear distance between side rails for all portable ladders must be 11.5 inches.
- The rungs and steps of portable metal ladders must be corrugated, knurled, dimpled, coated with skid-resistant material, or treated to minimize slipping.

Safe Usage of Ladders

- When portable ladders are used for access to an upper landing surface, the side rails must extend at least 3 feet above the upper landing surface. When such an extension is not possible, the ladder must be secured and a grasping device such as a grab rail must be provided to assist workers in mounting and dismounting the ladder. A ladder extension must not deflect under a load that would cause the ladder to slip off its support.

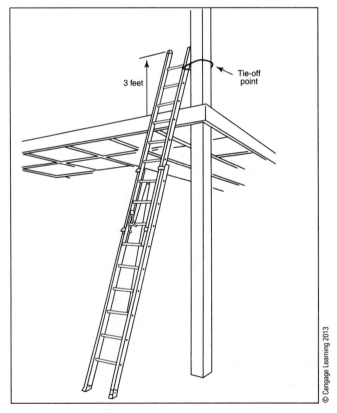

© Cengage Learning 2013

Ladder extended 3 feet over edge and secured properly

- Ladders must be maintained free of oil, grease, and other slipping hazards.

- Ladders must not be loaded beyond the maximum intended load for which they were built or beyond their manufacturer's rated capacity.

- Ladders must be used only for the purpose for which they were designed.

- Non-self-supporting ladders must be used at an angle where the horizontal distance from the top support to the foot of the ladder is approximately one-quarter of the working length of the ladder.

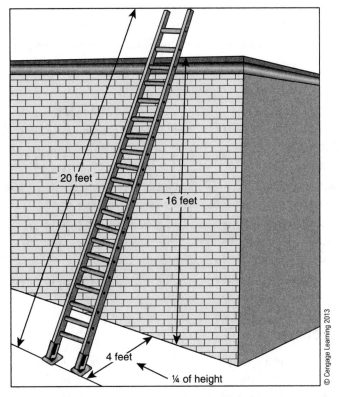

20 feet

16 feet

4 feet

¼ of height

© Cengage Learning 2013

Proper angle of extension ladder

- Fixed ladders must be used at a pitch no greater than 90 degrees from the horizontal, measured from the back side of the ladder.

- Ladders must be used only on stable and level surfaces unless secured to prevent accidental movement.

- Ladders must not be used on slippery surfaces unless secured or provided with slip-resistant feet to prevent accidental movement. Slip-resistant feet must not be used as a substitute for the care in placing, lashing, or holding a ladder upon slippery surfaces.

- Ladders placed in areas such as passageways, doorways, or driveways, or where they can be displaced by workplace activities or traffic, must be secured to prevent accidental movement or a barricade must be used to keep traffic or activities away from the ladder.

- The area around the top and bottom of the ladders must be kept clear.

- The top of a non-self-supporting ladder must be placed with two rails supported equally unless it is equipped with a single support attachment.

- Ladders must not be moved, shifted, or extended while in use.

- Ladders must have nonconductive side rails if they are used where the worker or the ladder could contact exposed energized electrical equipment.

- The top or top step of a stepladder must not be used as a step.

- Cross bracing on the rear section of stepladders must not be used for climbing unless the ladders are designed and provided with steps for climbing on both front and rear sections.

- Ladders must be inspected by a competent person for visible defects on a periodic basis and after any incident that could affect their safe use.

- Single-rail ladders must not be used.

- When ascending or descending a ladder, the worker must face the ladder.

- Each worker must use at least one hand to grasp the ladder when moving up or down the ladder.

- A worker on a ladder must not carry any object or load that could cause the worker to lose balance and fall.

Structural Defects

- Portable ladders with structural defects—such as broken or missing rungs, cleats, or steps, broken or split rails, corroded components, or other faulty or defective components—must immediately be marked defective or tagged with "Do Not Use" or similar language and withdrawn from service until repaired.

- Fixed ladders with structural defects—such as broken or missing rungs, cleats, or steps, broken or split rails, or corroded components—must be withdrawn from service until repaired.

- Defective fixed ladders are considered withdrawn from use when they are: (a) immediately tagged with "Do Not Use" or similar language; (b) marked in a manner that identifies them as defective; or (c) blocked (such as with a plywood attachment that spans several rungs).

- Ladder repairs must restore the ladder to a condition meeting its original design criteria before the ladder is returned to use.

Training Requirements

Under the provisions of the standard, employers must provide a training program for each employee using ladders and stairways. The program must enable each employee to recognize hazards related to ladders and stairways and to use proper procedures to minimize these hazards. For example, employers must ensure that each employee is trained by a competent person in the following areas, as applicable:

- The nature of fall hazards in the work area

- The correct procedures for erecting, maintaining, and disassembling the fall protection systems to be used

- The proper construction, use, placement, and care in handling of all stairways and ladders

- The maximum intended load-carrying capacities of ladders used. In addition, retraining must be provided for each employee, as necessary, so that the employee maintains the understanding and knowledge acquired through compliance with the standard

CHAPTER **8**

SAFETY INSPECTIONS

As you have read through this guide, you have learned about OSHA, how the agency was started, and how it conducts enforcement inspections. You also read about workers' compensation and learned how a good safety record can be a competitive advantage. The guide has also presented information on creating a safety and health program, conducting a job hazard analysis, and providing safety and health training. You then learned about OSHA regulations and safe work practices. The next step in creating a safe workplace for employees is to conduct safety inspections.

The safety inspection is designed to catch hazards missed at other stages. Inspections also allow you to monitor the safe work practices of employees and subcontractors. While an alert, trained, and competent workforce is the constant "real-time" protection against accident and injury, inspections provide the final clear and concentrated focus on potential problems. But let's be realistic. To gather a handful of people and tell them to walk the jobsite and "inspect" is virtually useless. The dentist who inspects your teeth at your semiannual visit has years of training and usually knows your history and what to look for. So does the mechanic who inspects your car. With this in mind, here are some things to consider about the inspection process.

Always know why an inspection is being conducted! In other words, make sure the objectives of the activity are known and clear to everyone involved. Some objectives could be to:

- Meet OSHA or other legal obligations
- Involve the team in safety
- Identify areas of undue risk and high loss potential
- Provide safety education
- Check past training and skill development
- Identify and develop positive safety attitudes
- Suggest better job methods
- Reinforce the positive efforts of people in the workplace

When it comes to OSHA, keep this fact in mind: OSHA only recommends general workplace inspections, but certain inspections are required. You should always check the standards to be sure you know what you must do at your jobsite. For example, the following items are generally necessary in most workplaces, but this isn't a complete list by any means. Check the standards for others not listed here.

- Cranes and derricks
- Industrial slings

- Manlifts
- Mechanical power presses and forging equipment
- Portable and fixed dry chemical extinguishers
- Powered industrial trucks
- Powered platforms
- Respiratory protection, including monthly inspections of emergency respirators
- Welding, cutting, and brazing equipment

IN-HOUSE SAFETY INSPECTIONS

Many companies find it helpful to use checklists for their inspections to ensure that important items are not overlooked. If they're helpful, that's fine, but your people should really be trained to question anything that doesn't appear safe or proper and not limit themselves to what's on the checklist. A sample safety inspection checklist can be found in Resource N, located at the back of this book and at dewalt.com/guides. This may be a good starting point for your company, but it is usually better to create a checklist specific to your company and your operations. When creating your own checklist, based it on:

- Past problems
- Standards that apply to your company
- Specific standards of concern to you
- Input from employees
- Your company safety and health program

When performing inspections, you could:

- Use inspections teams with broad safety skills
- Include supervisors, mechanics, and operators with specific backgrounds
- Have trainees make up checklists based on training completed
- Work from job procedures or JHAs

When putting checklists together, avoid excessive detail, vague criteria, and forms that try to impress or over-whelm. Remember, these are all just tools to aid in training. Once people become skilled inspectors, they won't need checklists and they probably won't let hazards sit until the inspection team cones by. Ideally, if the safety culture is strong, hazards will rarely crop up and most will be corrected on the spot by the first employee aware of the problem.

When it comes to documentation, you can write an inspection report on notebook paper, but a standard format and approach helps keep things organized. Here are some basic criteria for what to put on the report:

- Have a form that tells who, what, when, why, and where. You'll need this information to get the correction process working.
- Make the form or report easy to follow and use. Managers and those taking action on the report should be able to see at a glance the status of their organization.
- Include recommendations to give some guidance to those taking corrective action.
- Be helpful and encouraging. There is no need for an inspection to focus only on problems. If the team finds excellent conditions and positive safety behaviors, write it up! This is an opportunity for positive reinforcement!

- Rank findings and show the status of corrections so results can be tracked, plotted, and understood.
- Be sure all items are corrected! Inspections lose punch and management's safety credibility takes a real dive fast if results aren't obvious every time!

SCOPE OF SAFETY INSPECTION

The scope of your safety inspections should include the following:

Materials Handling and Storage	Equipment, job planning, layout, heights, floor loads, projection of materials, materials handling and storage methods, and training for material handling equipment
Building and Grounds Conditions	Floors, walls, ceilings, exits, stairs, walkways, ramps, platforms, driveways, and aisles
Housekeeping Program	Waste disposal, tools, objects, materials, leakage and spillage, cleaning methods, schedules, work areas, remote areas, and storage areas
Electricity	Equipment, switches, breakers, fuses, switch-boxes, junctions, special fixtures, circuits, insulation, extensions, tools, motors and grounding
Lighting	Type, intensity, controls, conditions, diffusion, location, and glare and shadow control
Heating and Ventilation	Type, effectiveness, temperature, humidity, controls, and natural and artificial ventilation and exhaust
Machinery	Points of operation, flywheels, gears, shafts, pulleys, key ways, belts, couplings, sprockets, chains, frames, controls, lighting for tools and equipment, brakes, exhausting, feeding, oiling, adjusting, maintenance, lockout/tagout, grounding, work space, location, and purchasing standards
Personnel	Experience training, including hazard identification training; methods of checking machines before use; type of clothing; personal protective equipment; use of guards; tool storage; work practices; and methods of cleaning, oiling, or adjusting machinery
Hand and Power Tools	Purchasing standards, inspection, storage, repair, types, maintenance, grounding, use, and handling
Chemicals	Storage, handling, transportation, spills, disposals, amounts used, labeling, toxicity or other harmful effects, warning signs, supervision, training, protective clothing and equipment, and hazard communication requirements
Fire Prevention	Extinguishers, alarms, sprinklers, smoking rules, exits, personnel assigned, separation of flammable materials and dangerous operations, explosive-proof fixtures in hazardous locations, and waste disposal
Maintenance (including tracking and abatement of preventive and regular maintenance)	Regularity, effectiveness, training of personnel, materials and equipment used, records maintained, method of locking out machinery, and general methods

Personal Protective Equipment	Type, size, maintenance, repair, storage, assignment of responsibility, purchasing methods, standards observed, training in care and use, rules of use, and method of assignment
Transportation	Motor vehicle safety, seat belts, vehicle maintenance, and safe driver programs
Review	Emergency action plans, equipment, and training

THIRD-PARTY SAFETY INSPECTIONS

Safety inspections conducted by a third party can provide an unbiased look at operations and work practices. There are a few sources for these third-party safety inspections:

- **Insurance Companies:** Insurance companies will often inspect their clients periodically to help them with their safety and health program. You may have already had an insurance company inspection, but if not, ask for one. This may be a complementary service that will benefit your construction company.

- **Safety Consultants:** Safety consultants can assist companies with safety inspections on a more regular basis. These inspections will often improve compliance and help a construction company achieve a better safety record.

- **OSHA Consultation Services:** OSHA's On-site Consultation Program offers free and confidential advice to small and medium-sized businesses in all states across the country, with priority given to high-hazard worksites. In FY 2010, responding to requests from small employers looking to create or improve their safety and health management systems, OSHA's On-site Consultation Program conducted over 30,000 visits to small-business worksites covering over 1.5 million workers across the nation.

On-site consultation services are separate from enforcement and do not result in penalties or citations. Consultants from state agencies or universities work with employers to identify workplace hazards, provide advice on compliance with OSHA standards, and assist in establishing safety and health management systems. For more information on these services from OSHA, go to http://www.osha.gov and visit their small business consultation section.

CHAPTER 9

FUNDAMENTALS OF ACCIDENT AND INCIDENT INVESTIGATIONS

As discussed in Chapter 2, recordkeeping is a critical part of a construction company's safety and health efforts for several reasons:

- Keeping track of work-related injuries and illnesses can help the company prevent them in the future.
- Using injury and illness data helps identify problem areas. The more a company knows, the better it can identify and correct hazardous workplace conditions.
- Construction companies can better administer their safety and health programs with accurate records.

It is also critically important to investigate all accidents and incidents that occur at the jobsite. By investigating, a company will be better equipped to find the root cause of the accident or incident. Understanding the root cause will assist companies to make continued improvements in their safety and health programs.

ACCIDENT VERSUS INCIDENT

Investigating accidents is a no-brainer. A company will want to find out what happened, discover contributing factors, and determine the cause so it can prevent a reoccurrence of the accident. But why should companies also investigate incidents? Let us start by defining each term:

- **Accident:** An accident is defined as an undesirable event that occurs unintentionally and usually results in harm, injury, damage, or loss.
- **Incident:** An incident is defined as an individual occurrence or event that is likely to take place again if all the elements leading up to the incident remain the same. An incident is also referred to as a near miss.

The main difference between an accident and an incident is that an accident results in a loss and an incident does not. The term "near miss" can also be used to describe an incident. Many companies understand that the accident caused the injury. However, a similar event could occur but not result in an accident.

As an example, a forklift overturned and crushed the leg of the operator. The construction company's first-aid provider responded to the scene, and once the paramedics left with the injured forklift operator, workers standing around the jobsite discussed what they had seen. A couple of workers told stories of the same forklift operator driving the forklift too fast on the jobsite. Other workers told of seeing the operator driving the forklift and not wearing his seatbelt. One worker spoke about a time when the forklift almost rolled because of operator error.

The company wished it had heard that information prior to the accident. It could have retrained the operator and taken other steps to prevent those incidents from recurring. Instead, it had an employee who lost his leg in an accident. If there had been a system for investigating all incidents, a major accident could have been prevented.

ACCIDENT AND INCIDENT INVESTIGATIONS

When to Conduct the Investigation

It is important that any investigation occur as soon as possible. The less time between an accident or incident and the investigation, the more accurate the information that can be obtained.

Goal of the Investigation

A key element in the investigation is examining the causes and results of any accident or incident without prejudice. The investigator must begin the investigation with an open mind. No hypothesis should be made, and any conclusion should be based on information that is known to be full and accurate.

It is helpful for the investigator to ask open-ended questions and not put words into witnesses' mouths. The investigator should not blame people but rather emphasize the significance of seeking the reasons for the incident to prevent a reoccurrence.

Attempts to change people and their behaviors are far less likely to succeed than making changes in their surroundings so that the consequences of a mistake are either eradicated or reduced. Instead of an emotional discussion about patterns of behavior, a more constructive approach is to attempt to modify the environment. It is more effective to adjust the circumstances producing a mistake than to attempt to change human nature.

As an example, if construction materials are poorly stacked and they collapse, causing an injury to a worker nearby, the apparent cause could be poor materials handling practices. However, possible contributing factors could include employees not understanding the hazard of their actions, the storage area being unsuitable for the task, or the materials being inadequately maintained. Thus, the true basic causes could identify the need in this case for:

- Further worker training
- Better planning and layout of the storage area
- New equipment or techniques

A thorough accident or incident investigation may require photographs, drawings, and/or technical expertise before the final causes of an accident or incident can be determined and satisfactory controls are chosen.

Basic Steps of an Investigation

When initiating the investigation, take the following steps:

- If workers have been injured, make sure they are given appropriate medical attention without delay.
- Control the scene. Place barriers, caution tape, or other devices to prevent others from disturbing the scene.
- Start the investigation as quickly as possible. Conduct interviews at the scene if possible. Ensure that the witnesses discuss the accident or incident in relative privacy so they feel comfortable. Begin with the witnesses who can contribute the most information. These are usually the witnesses who were closest to the accident or incident.

- Take notes during each interview. After the interview, repeat the witness's statement to ensure that you have understood correctly. Document this on a witness statement form and ask the witness to sign the form.
- Close each interview on a positive note and thank all witnesses for their assistance.
- Take immediate corrective action where reasonable to help prevent additional hazards.
- Complete the investigation report with recommendations and follow-up actions.
- Ensure follow-up action occurs.

Key Questions

- *Who?* Get the names of everyone involved, near, present, or aware of possible contributing factors.
- *What?* Describe materials and equipment that were involved, check for defects, get a precise description of chemicals involved, and so on.
- *Where?* Describe the exact location; note all relevant facts (lighting, weather, ground conditions).
- *When?* Note the exact time, date, and other factors (shift change, work cycle, break period).
- *How?* Describe the usual sequence of events and the actual sequence of events before, during, and after the accident or incident.
- *Why?* Find all possible direct and indirect causes and how to keep the event from happening again.

Essential Notions

Causes of accidents or incidents are rarely simple when they are investigated closely. Behind every case there are many contributing factors and causes. The key is to identify those that can be most effectively acted upon to prevent recurrences.

Investigations should emphasize the long-term eradication of injury, loss, or damage. The focus should be on systems deficiencies in preference to human factors.

After identifying causes and factors, suitable improvement actions must be identified and implemented. Several forms, such as those shown on the following pages, can help the investigator with this task.

ACCIDENT / INCIDENT INVESTIGATION REPORT

COMPANY: _____   DATE: _____

ADDRESS: _____

JOBSITE NAME: _____

ADDRESS: _____

NAME OF INJURED	DATE OF ACCIDENT/INCIDENT	TIME OF ACCIDENT/INCIDENT
HOME ADDRESS AND PHONE	EMPLOYEE'S USUAL OCCUPATION	OCCUPATION AT TIME OF ACCIDENT/INCIDENT

EMPLOYMENT CATEGORY	LENGTH OF EMPLOYMENT	TIME in OCCUPATION
❑ Regular, full-time ❑ Seasonal ❑ Temporary ❑ Regular, part-time ❑ Nonemployee	❑ Less than 1 mo. ❑ 6 mos. to 5 yrs. ❑ 1–5 mos. ❑ > 5 years	❑ Less than 1 mo. ❑ 6 mos. to 5 yrs. ❑ 1–5 mos. ❑ > 5 years

NAMES OF OTHER INJURED IN SAME ACCIDENT/INCIDENT

NATURE of INJURY and PART of BODY	TIME of INJURY	SEVERITY of INJURY
	A. _____ A.M/P.M B. Time within shift C. Type of Shift	❑ Fatality ❑ Medical Treatment ❑ First Aid ❑ Other, specify _____

TASK and ACTIVITY at TIME of ACCIDENT/INCIDENT	SUPERVISION at TIME of ACCIDENT/INCIDENT
A. General type of task B. Specific Activity C. Employee was working: ❑ Alone ❑ With crew or fellow worker ❑ Other, specify	❑ Directly Supervised ❑ Not Supervised ❑ Indirectly Supervised ❑ Supervision not feasible

LOCATION OF ACCIDENT/ INCIDENT	PHASE OF EMPLOYEES WORKDAY AT TIME OF ACCIDENT/INCIDENT	WEATHER CONDITIONS AT TIME OF ACCIDENT/INCIDENT
     ON EMPLOYER'S PREMISES? ❑ Yes ❑ No	❑ During rest period ❑ Performing work duties ❑ During meal period ❑ Entering worksite ❑ Working overtime ❑ Leaving worksite ❑ Other, specify _____	

NAMES OF WITNESS TO THE ACCIDENT/INCIDENT

DESCRIBE HOW THE ACCIDENT/INCIDENT OCCURRED.

ACCIDENT SEQUENCE: Describe in reverse order of occurrence events preceding the injury and accident. Starting with the injury and moving backward in time, reconstruct the sequence of events that led to the injury.

A. Injury Event _____

B. Accident Event _____

C. Preceding Event #1 _____

D. Preceding Event #2. #3. etc. _____

CASUAL FACTORS. Events and conditions that contributed to the accident. Be sure and describe in detail if the proper safety equipment was being used and if it was used correctly.

CORRECTIVE ACTIONS. Those that have been, or will be, taken to prevent recurrence.

Investigation Officer _____

Company _____

Signature   Date

Interpreter _____

Company _____

Signature   Date

Sample accident/incident investigation report

WITNESS STATEMENT FORM

WITNESS NAME: _____ WITNESS EMPLOYER: _____

ADDRESS: _____

PHONE: _____

AGE: _____ OCCUPATION: _____

BRIEF DESCRIPTION OF ACCIDENT/INCIDENT

RELATIONSHIP TO INJURED PARTY

Immediately before the accident, what did you see? Did you notice the injured employee doing anything wrong? Did you warn them? Where were you at? How far away? What did you see?

During the accident, what did you see?

Immediately after the accident, what did you see?

Have you spoken with anyone else concerning this incident?

Additional Comments:

_____ _____
Witness   Date   Investigator   Date

_____
Interpreter   Date

Courtesy of SFI Compliance Inc.

if you run out of room, use the back of this page

Sample witness statement form

CHAPTER 10

WHAT TO DO DURING AN OSHA INSPECTION

Chapter 2 presented information on how OSHA conducts inspections. What do you do if OSHA arrives on your site? The following information will assist construction companies that experience an OSHA enforcement inspection:

Arrival of the Compliance Officer	• Verify the compliance officer's credentials.
	
	 • Determine why the compliance officer wants to inspect your project; for example: complaint, accident, programmed inspection, imminent danger, follow-up.
	   ○ If it is a complaint inspection, ask to have a copy of the complaint.
	   ○ If it is a programmed inspection, ask for a focused inspection.
	
	 • Call your safety manager immediately. Brief your safety manager on the situation, and then ask the compliance officer if the officer will speak to your safety manager. Your safety manager should attempt to be present during the inspection and will let the compliance officer know when the manager can be on site and ask if the officer can wait to begin the inspection until the manager arrives. Many compliance officers will wait a reasonable time for management to arrive on-site prior to starting an inspection.

Opening Conference	• Request an opening conference if the compliance officer does not call for one. • Have management present for the inspections: the project superintendent and/or the project manager. • Take detailed notes of everything discussed. • Keep all publications and documents given to you by the compliance officer and note who gave it to you and the date of its receipt. • If more than one compliance officer is involved, find out if they plan to make the inspection in one group or split into two or more. If they want to divide into two or more groups, tell the compliance officer you will have an employer representative with each group.

(continues)

	• Be cooperative with the compliance officer. At all times prior to, during, and after the inspection, act in a professional, businesslike manner. Never enter into personal arguments with the compliance officer. • Tell the compliance officer that you expect him to advise the company of all suspected violations and the standard involved. Inform the compliance officer that you will be taking notes of all the suspected violations he informs you of so that there will be no dispute as to whether the company was informed.
Company Records	• Do not volunteer to give or let the compliance officer look at any safety inspection reports made by the safety department personnel, insurance carrier, etc., other than those reports required by OSHA. This includes not giving or letting the compliance officer look at any such reports that have been made on subcontractors' activities. Let the compliance officer know that you will request that these documents be sent to them.
During the Inspection	General • Have the designated employer representative accompany the compliance officer at all times. Never allow the compliance officer to have free and unlimited access to your work without an employer representative. • Control the inspection. Treat the compliance officer as you would a guest in your house; he is there with permission and will be expected to follow all instructions given him and will be required to conduct the inspection in such a manner that it does not disrupt the scheduled work. Remember, it is the company's construction project, not OSHA's. The compliance officer is to be treated as any other visitor—under your control while on your project. • Tell him you need to take your own notes and pictures and gather information at the same time. • Take detailed notes of everything seen, discussed, and done by the compliance officer. • Take photographs of everything the compliance officer photographs. If the equipment, work area, etc., can be photographed from a more favorable position (different angle, greater distance, etc.), photograph it from the different position. • Do not allow any employee to perform demonstrations for the compliance officer. Example: If a truck is idle and the compliance officer asks you or an employee to operate it so he can see if the horn or backup alarm is working, refuse to do it. Only allow the compliance officer to see the work as it is normally being done. • Do not volunteer any information. The compliance officer is trained to obtain admissions from companies. Be careful answering questions. When in doubt, ask the officer to restate the question. • Always tell the truth.

	• Do not issue orders, such as "clean up the trash," during the inspection to correct conditions that the compliance officer has not noticed. The immediate abatement of an alleged violation will not preclude being cited by the compliance officer and may alert him to the condition. • Make sure you fully understand everything the compliance officer does or comments on. If you do not understand everything he says and does, ask him questions. Insist on having time to record all facts, even if the inspection is slowed down. Employee Interviews • The compliance officer may interview any employee privately. • For interviews held with employees in your presence, record the names and companies of all employees interviewed. Record the content of the conversations with the employees. • You have the right to be in attendance with management personnel.
Closing Conference	• Have the designated employer representative and the project superintendent present. • Take notes of everything discussed and record what documents were distributed by the compliance officer. • Make sure that no questions you have concerning the inspection go unanswered. If the compliance officer feels that violations exist on the project, find out exactly why he feels that way. Tell him that you are noting every potential violation that he advises may exist so that there will be no dispute at a later conference or hearing. Hand the list back and ask if it includes everything and what other items may exist. If he is unsure or does not list any others, tell him that you are noting that no additional items were included in response to your questions. • Don't give any estimates of abatement time needed to correct any alleged violations.
After the Inspection	The compliance officer will usually inform you at the closing conference whether or not you will receive a citation. OSHA has up to 6 months to issue citations. If any violations were pointed out by OSHA during the inspection, make sure to correct those violations and document your correction activity.
Receipt of Citation	If you receive a citation from the OSHA inspection, you will need to post the citation in a place where employees can view it. The citation must be posted for 3 days or until the violation(s) is/are corrected, whichever is greater. You have a right to schedule an informal conference with OSHA. This will allow you to tell your side and possibly mitigate the citation. All construction companies should take advantage of the informal conference with OSHA.

CHAPTER 11

EMERGENCY ACTION PLAN

Employers and employees at all jobsites should take steps to prepare for an emergency. Being prepared can result in a much better outcome. This guide has presented a lot of information on OSHA and safe work practices. Construction companies can control most hazards by following OSHA regulations and maintaining safe work practices. Emergencies that are out of the control of a company should also be addressed. They should be planned for so that employers and their employees can safely respond at the jobsite. Many resources are available to assist employers in preparing an emergency action plan. The following checklist and much more can be found at http://www.osha.gov. This checklist is also found at dewalt.com/guides.

EMERGENCY ACTION PLAN CHECKLIST[1]

General Issues		
☐	Does the plan consider all natural or man-made emergencies that could disrupt the workplace?	Common sources of emergencies identified in emergency action plans include fires, explosions, floods, hurricanes, tornadoes, releases of toxic material, radiological and biological accidents, civil disturbances, and workplace violence.
☐	Does the plan consider all potential internal sources of emergencies that could disrupt the workplace?	Conduct a hazard assessment of the workplace to identify any physical or chemical hazards that may exist and could cause an emergency.
☐	Does the plan consider the impact of these internal and external emergencies on the workplace's operations and is the response tailored to the workplace?	Brainstorm worst-case scenarios by asking yourself what you would do and what would be the likely impact on your operation, then devise appropriate responses.
☐	Does the plan contain a list of key personnel with contact information as well as contact information for local emergency responders, agencies, and contractors?	Keep the list of key contacts current and make provisions for an emergency communications system such as a cellular phone, a portable radio unit, or other means so that contact with local law enforcement, the fire department, and others can be swift.

(continues)

---

[1] Occupational Safety & Health Administration, U.S. Department of Labor. *Evacuation Plans and Procedures*, http://www.osha.gov.

☐	Does the plan contain the names, titles, departments, and telephone numbers of individuals to contact for additional information or an explanation of duties and responsibilities under the plan?	List names and contact information for individuals responsible for implementation of the plan.
☐	Does the plan address how rescue operations will be performed?	Unless you are a large employer handling hazardous materials and processes or have employees regularly working in hazardous situations, you will probably choose to rely on local public resources, such as the fire department, which is trained, equipped, and certified to conduct rescues. Make sure any external department or agency identified in your plan is prepared to respond as outlined in the plan. Untrained individuals may endanger themselves and those they are trying to rescue.
☐	Does the plan address how medical assistance will be provided?	Most small employers do not have a formal internal medical program and make arrangements with medical clinics or facilities close by to handle emergencies. If an infirmary, clinic, or hospital is not close to the workplace, ensure that on-site person(s) have adequate training in first aid. The American Red Cross, some insurance providers, local safety councils, fire departments, or other resources may be able to provide this training. Treatment of a serious injury should begin within 3 to 4 minutes of the accident. Consult with a physician to order appropriate first-aid supplies for emergencies. Establish a relationship with a local ambulance service so that transportation is readily available for emergencies.
☐	Does the plan identify how or where personal information on employees can be obtained in an emergency?	In the event of an emergency, it could be important to have ready access to personal information about employees. This includes their home telephone numbers, the names and telephone numbers of their next of kin, and medical information.
Evacuation Policy and Procedure		
☐	Does the plan identify the conditions under which an evacuation would be necessary?	The plan should identify the different types of situations that will require an evacuation of the workplace. This might include a fire, earthquake, or chemical spill. The extent of evacuation may be different for different types of hazards.

❐	Does the plan identify a clear chain of command and designate a person authorized to order an evacuation or shutdown of operations?	It is common practice to select a responsible individual to lead and coordinate an emergency plan and evacuation. It is critical that employees know who the coordinator is and understand that this person has the authority to make decisions during emergencies. The coordinator should be responsible for assessing the situation to determine whether an emergency exists requiring activation of the emergency procedures, overseeing emergency procedures, notifying and coordinating with outside emergency services, and directing shutdown of utilities or plant operations if necessary.
❐	Does the plan address the types of actions expected of different employees for the various types of potential emergencies?	The plan may specify different actions for employees depending on the emergency. For example, employers may want to have employees assemble in one area of the workplace if it is threatened by a tornado or earthquake but evacuate to an exterior location during a fire.
❐	Does the plan designate who, if anyone, will stay to shut down critical operations during an evacuation?	You may want to include in your plan locations where utilities (such as electricity and gas) can be shut down for all or part of the facility. All individuals remaining behind to shut down critical systems or utilities must be capable of recognizing when to abandon the operation or task and evacuate themselves.
❐	Does the plan outline specific evacuation routes and exits and are these posted in the workplace where they are easily accessible to all employees?	Most employers create maps from floor diagrams with arrows that designate the exit route assignments. These maps should include locations of exits, assembly points, and equipment (such as fire extinguishers, first-aid kits, spill kits) that may be needed in an emergency. Exit routes should be clearly marked and well lit, wide enough to accommodate the number of evacuating personnel, unobstructed and clear of debris at all times, and unlikely to expose evacuating personnel to additional hazards.

(continues)

❑	Does the plan address procedures for assisting people during evacuations, particularly those with disabilities or who do not speak English?	Many employers designate individuals as evacuation wardens to help move employees from danger to safe areas during an emergency. Generally, 1 warden for every 20 employees should be adequate, and the appropriate number of wardens should be available at all times during working hours. Wardens may be responsible for checking offices and bathrooms before being the last person to exit an area as well as ensuring that fire doors are closed when exiting. Employees designated to assist in emergency evacuation procedures should be trained in the complete workplace layout and various alternative escape routes. Employees designated to assist in emergencies should be made aware of employees with special needs (who may require extra assistance during an evacuation), how to use the buddy system, and any hazardous areas to avoid during an emergency evacuation.
❑	Does the plan identify one or more assembly areas (as necessary for different types of emergencies) where employees will gather and a method for accounting for all employees?	Accounting for all employees following an evacuation is critical. Confusion in the assembly areas can lead to delays in rescuing anyone trapped in the building or unnecessary and dangerous search-and-rescue operations. To ensure the fastest, most accurate accounting of employees, consider taking a head count after the evacuation. The names and last known locations of anyone not accounted for should be passed on to the official in charge.
❑	Does the plan address how visitors will be assisted in evacuation and accounted for?	Some employers have all visitors and contractors sign in when entering the workplace. The hosts and/or area wardens, if established, are often tasked with assisting these individuals to evacuate safely.
Reporting Emergencies and Alerting Employees in an Emergency		
❑	Does the plan identify a preferred method for reporting fires and other emergencies?	Dialing 911 is a common method for reporting emergencies if external responders are utilized. Internal numbers may be used as well. Internal numbers are sometimes connected to intercom systems so that coded announcements may be made. In some cases, employees are requested to activate manual pull stations or other alarm systems.

☐	Does the plan describe the method to be used to alert employees, including disabled workers, to evacuate or take other action?	Make sure alarms are distinctive and recognized by all employees as a signal to evacuate the work area or perform other actions identified in the plan. Sequences of horn blows or different types of alarms (bells, horns, etc.) can be used to signal different responses or actions from employees. Consider making available an emergency communications system, such as a public address system, for broadcasting emergency information to employees. Ideally, alarms can be heard, seen, or otherwise perceived by everyone in the workplace, including those who may be blind or deaf. Otherwise, floor wardens or others must be tasked with ensuring all employees are notified. You might want to consider providing an auxiliary power supply in the event of an electrical failure.
Employee Training and Drills		
☐	Does the plan identify how and when employees will be trained so that they understand the types of emergencies that may occur, their responsibilities, and actions as outlined in the plan?	Training should be offered to employees when the initial plan is developed and when new employees are hired. Employees should be retrained when the plan changes due to a change in the layout or design of the facility; when new equipment, hazardous materials, or processes are introduced that affect evacuation route; or when new types of hazards are introduced that require special actions. General training for employees should address the following: • Individual roles and responsibilities; • Threats, hazards, and protective actions; • Notification, warning, and communications procedures; • Emergency response procedures; • Evacuation, shelter, and accountability procedures; • Location and use of common emergency equipment; and • Emergency shutdown procedures. You may also need to provide additional training to employees (i.e. first-aid procedures, portable fire extinguisher use, etc.), depending on the responsibilities allocated to employees in the plan.

(continues)

❏	Does the plan address how and when retraining will be conducted?	If training is not reinforced, it will be forgotten. Consider retraining employees annually.
❏	Does the plan address if and how often drills will be conducted?	Once the emergency action plan has been reviewed with employees and everyone has had the proper training, it is a good idea to hold practice drills as often as necessary to keep employees prepared. Include outside resources such as fire and police departments when possible. After each drill, gather management and employees to evaluate the effectiveness of the drill. Identify the strengths and weaknesses of the plan and work to improve it.

CHAPTER 12

SAFETY MANAGEMENT SYSTEMS

This guide has offered many options to create and make safety and health programs more effective. Safety management is a comprehensive business management system designed to manage occupational safety and health elements in the workplace. Safety management professionals provide a methodical way for identifying hazards and performing risk assessment. Safety management professionals help their organization control risks.

Safety management systems are used by proactive companies to help avoid workplace injuries and illnesses. Construction companies can benefit greatly from advanced safety management systems because construction is considered a high-hazard industry. Through effective safety management systems, construction companies can protect their employees from injury and their company from losses.

SAFETY MADE SIMPLE

A safety management system should match the organization. Construction companies that do not have large safety departments may find developing a methodical way to manage safety extremely cumbersome. However, safety must still be managed, even through a simpler, but still effective, system. The steps in the following safety management system can be used by such companies:

- Create safety and health program, including an emergency action plan
- Prepare JHAs for all tasks workers will encounter
- Train workers in the hazards they may face
  - Retrain them as necessary
  - Conduct safety meetings regularly—weekly is advised
- Conduct jobsite safety inspections
- Investigate all incidents
- Evaluate the program and constantly make adjustments

Following a safety management system such as this one can give you that competitive advantage that you desire.

DOCUMENTATION

"Not written, not done"

With all safety efforts, make sure the activities are documented. They will have a definite impact on jobsite safety. However, construction companies also need to take steps to protect their interests. Imagine having to defend yourself in a court of law—having written documentation for every effort you have made would be important. Don't waste efforts by not being able to prove them.

SAFETY CULTURE

It has been observed at the OSHA VPP sites and confirmed by independent research that of any process developing strong safety cultures has the single greatest impact on accident reduction. It is for this single reason that developing these cultures should be top priority for all managers and supervisors.

What Is a Safety Culture—How Will It Impact My Company?

Safety cultures consist of a company's shared beliefs, practices, and attitudes. Culture is the atmosphere created by those beliefs, attitudes, etc., which shape our behavior. An organization's safety culture is the result of a number of factors, such as:

- Management and employee norms, assumptions, and beliefs
- Management and employee attitudes
- Values, myths, stories
- Policies and procedures
- Supervisor priorities, responsibilities, and accountability
- Production and bottom-line pressures versus quality issues
- Actions or lack of actions to correct unsafe behaviors
- Employee training and motivation
- Employee involvement or "buy-in"

In a strong safety culture, everyone feels responsible for safety and pursues it on a daily basis; employees go beyond "the call of duty" to identify unsafe conditions and behaviors and intervene to correct them. For instance, in a strong safety culture any worker would feel comfortable walking up to the plant manager or CEO and reminding him or her to wear safety glasses. This type of behavior would not be viewed as forward or overzealous, but would be valued by the organization and rewarded. Likewise, co-workers routinely look out for one another and point out unsafe behaviors to each other.

A company with a strong safety culture typically experiences few at-risk behaviors; consequently, they also experience low accident rates, low turnover, low absenteeism, and high productivity. They are usually companies who are extremely successful by excelling in all aspects of business and operations.

Creating a safety culture takes time. It is frequently a multi-year process. A series of continuous process improvement steps can be followed to create a safety culture. Employer and employee commitment are hallmarks of a true safety culture where safety is an integral part of daily operations. A company at the beginning of the road toward developing a safety culture may exhibit a level of safety awareness, consisting of safety posters and warning signs. As more time and commitment are devoted, a company will begin to address physical hazards and may develop safety recognition programs, create safety committees, and start incentive programs.

Top management support of a safety culture often results in acquiring a safety director, providing resources for accident investigations, and safety training. Further progress toward a true safety culture uses accountability systems. These systems establish safety goals, measure safety activities, and charge costs back to the units that incur them. Ultimately, safety becomes everyone's responsibility, not just the safety director's. Safety becomes a value of the organization and is an integral part of operations. Management and employees are committed and involved in preventing losses.

Over time, the norms and beliefs of the organization shift the focus from eliminating hazards to eliminating unsafe behaviors and building systems that proactively improve safety and health conditions. Employee safety and doing something the right way takes precedence over short-term production pressures. Simultaneously, production does not suffer but is enhanced due to the level of excellence developed within the organization.

Building a Safety Culture

Any process that brings all levels of the organization together to work on a common goal that everyone holds in high value will strengthen the organizational culture. A unique process that can do this is worker safety and health. It is one of the few initiatives that offer significant benefits for the frontline workforce. As a result, buy-in can be achieved, enabling the organization to effectively implement change. Obtaining frontline buy-in for improving worker safety and health is much easier than getting buy-in for improving quality or increasing profitability. When the needed process improvements are implemented, all three areas—safety and health, quality, and profitability—typically improve and a culture develops that supports continuous improvement in all areas.

The following sections discuss the major processes and milestones that are required to successfully implement a change process for safety and health. It intentionally focuses on the process rather than on individual tasks. The tendency to focus on the accomplishment of tasks, i.e., to train everyone on a particular concern or topic or implement a new procedure for incident investigations, etc., is common. Companies that maintain their focus on the larger process to be followed are far more successful. They can distinguish the "forest" from the "trees" and thus can make mid-course adjustments as needed. They never lose sight of their intended goals; therefore, they tend not to get distracted or allow obstacles to interfere with their mission. The process itself will take care of the task implementation and ensure that the appropriate resources are provided and priorities are set.

Management Processes Typically Ripe for Improvement

- Define safety responsibilities for all levels of the organization, e.g., safety is a line management function.
- Develop upstream measures, e.g., number of reports of hazards/suggestions, number of committee projects/successes, etc.
- Align management and supervisors through establishing a shared vision of safety and health goals and objectives versus production.
- Implement a process that holds managers and supervisors accountable for visibly being involved, setting the proper example, and leading a positive change for safety and health.
- Evaluate and rebuild any incentives and disciplinary systems for safety and health as necessary.
- Ensure the safety committee is functioning appropriately, e.g., membership, responsibilities/functions, authority, meeting management skills, etc.
- Provide multiple paths for employees to bring suggestions, concerns, or problems forward. One mechanism should use the chain of command and ensure no repercussions. Hold supervisors and middle managers accountable for being responsive.
- Develop a system that tracks and ensures the timeliness of hazard correction. Many jobsites have been successful in building this within an existing work order system.
- Ensure reporting of injuries, first-aid incidents, and near misses. Educate employees on the accident pyramid and the importance of reporting minor incidents. Prepare management for an initial increase in

incidents and rise in rates. This will occur if underreporting has existed in the organization. The rates will level off, then decline as the system changes take hold.

- Evaluate and rebuild the incident investigation system as necessary to ensure that it is timely, complete, and effective. It identify the root causes and avoid blaming workers.

Obtain Top Management Buy-in

This is the very first step that needs to be accomplished. Top managers must be on board. If they are not, safety and health concerns will compete with core business issues such as production and profitability, a battle that safety and health will almost always lose. Top management must understand the need for change and be willing to support it. Demonstrating the costs to the organization in terms of dollars (direct and indirect costs of accidents) lost and the organizational costs (fear, lack of trust, feeling of being used, etc.) can be compelling in terms of convincing management that it needs to do something different. Losses due to accidents are bottom-line costs to the organization; controlling them will more than pay for the needed changes. In addition, when these arguments are successful, they will also go a long way in eliminating organizational barriers such as fear, lack of trust, etc.—issues that typically get in the way of everything that the organization wants to do.

Continue Building Buy-in

The needed changes are accomplished by building an alliance or partnership between management, the union (if one exists), and employees. A compelling reason for the changes must be spelled out to everyone. People have to understand WHY they are being asked to change what they normally do and what it will look like if they are successful. This needs to be done up front. If people get wind that something "is going down" and haven't been formally told anything, they naturally tend to resist and opt out.

Build Trust

Trust is a critical part of accepting change, and management needs to know that this is part of the bigger picture, outside of all the details. Trust will occur as different levels within the organization work together and begin to see success.

Conduct Self-Assessments/Benchmarking

To get where you want to go, you must know from where you are starting. A variety of self-audit mechanisms can be employed to compare jobsite processes with other recognized models of excellence, such as OSHA Star VPP sites. Visiting other sites to gain first-hand information is also invaluable.

Initial Training

Conduct initial training of management/supervisory staff, union leadership (if present), safety and health committee members, and a representative number of hourly employees. Training may include both safety and health training and any needed management, team building, hazard recognition, or communication training, etc. This creates a core group of people to draw upon as resources and also gets key personnel onboard with needed changes.

Establish a Steering Committee

The steering committee should be comprised of management, employees, union (if one exists), and safety staff. The purpose of this group is to facilitate, support, and direct the change processes. This will provide overall guidance and direction and avoid duplication of effort. To be effective, the group must have the authority to get things done.

Develop Site Safety Vision

The safety vision should include key policies, goals, measures, and strategic and operational plans. These policies provide guidance and serve as a check that can be used to ask if the decision about to be made supports or detracts from the intended safety and health improvement process.

Align the Organization

Establish a shared vision of safety and health goals and objectives versus production. Upper management must be willing to support it by providing resources (time) and holding managers and supervisors accountable for doing the same. The entire management and supervisory staff need to set the example and lead the change. It's more about leadership than management.

Define Specific Roles

Roles and responsibilities for safety and health should be defined at all levels of the organization. Safety and health must be viewed as everyone's responsibility. How the organization is to deal with competing pressures and priorities, i.e., production versus safety and health, needs to be clearly spelled out.

Develop a System of Accountability

This should be done for all levels of the organization. Everyone must play by the same rules and be held accountable for their areas of responsibility. A sign of a strong culture is individuals holding themselves accountable.

Develop a Measurement System

There should be an ongoing measurement and feedback system. Upstream activity measures that encourages positive change should drive the system. Examples include the number of hazards reported or corrected, number of inspections, number of equipment checks, JHAs, pre-start-up reviews conducted, etc.

While it is always nice to know what the bottom-line performance is, i.e., accident rates, overemphasis on rates and using them to drive the system typically only drives accident reporting under the table. It is all too easy to manipulate accident rates, which will only result in risk issues remaining unresolved and the probability of more serious events occurring in the future.

Develop Policies for Recognition

Recognition polices, rewards, incentives, and ceremonies can have a great impact. Again, employees should be rewarded for doing the right things and be encouraged to participate in upstream activities. Continually reevaluate these policies to ensure their effectiveness and to ensure that they do not become entitlement programs.

Awareness Training and Kick-off

It's not enough for only part of the organization to be involved and know about the change effort—the entire jobsite needs to know and be involved in some manner. A kick-off celebration can be used to announce that it's a "new day" and to seek buy-in for any new procedures and programs.

Implement Process Changes

Process changes must include the involvement of management, the union (if one is present), and employees using the "Plan to Act" process of Total Quality Management (TQM).

Continually Measure Performance, Communicate Results, and Celebrate Successes

Publicizing results is very important to sustaining efforts and keeping everyone motivated. Everyone needs to be updated throughout the process. Progress reports given during normal shift meetings allow time for comments to be given to the steering committee, open communications, and also allow for input. Everyone needs to have a voice; otherwise, they will be reluctant to buy in. A system can be as simple as using current meetings, a bulletin board, and a comment box.

Ongoing Support

Reinforcement, feedback, reassessment, mid-course corrections, and ongoing training are vital to sustaining continuous improvement.

OSHA RECORDKEEPING FORMS

The following pages contain some of the recordkeeping forms as required by OSHA. These forms can also be found at dewalt.com/guides. The full forms can be found at http://www.osha.gov.

Courtesy of the Occupational Safety & Health Administration, U.S. Department of Labor

OSHA's Form 300 (Rev. 01/2004)

Log of Work-Related Injuries and Illnesses

You must record information about every work-related death and about every work-related injury or illness that involves loss of consciousness, restricted work activity or job transfer, days away from work, or medical treatment beyond first aid. You must also record significant work-related injuries and illnesses that are diagnosed by a physician or licensed health care professional. You must also record work-related injuries and illnesses that meet any of the specific recording criteria listed in 29 CFR Part 1904.8 through 1904.12. Feel free to use two lines for a single case if you need to. You must complete an Injury and Illness Incident Report (OSHA Form 301) or equivalent form for each injury or illness recorded on this form. If you're not sure whether a case is recordable, call your local OSHA office for help.

Attention: This form contains information relating to employee health and must be used in a manner that protects the confidentiality of employees to the extent possible while the information is being used for occupational safety and health purposes.

Year 20 ____

U.S. Department of Labor
Occupational Safety and Health Administration

Form approved OMB no. 1218-0176

Establishment name _____

City _____ State _____

Identify the person

(A) Case no.	(B) Employee's name	(C) Job title (e.g., Welder)

Describe the case

(D) Date of injury or onset of illness	(E) Where the event occurred (e.g., Loading dock north end)	(F) Describe injury or illness, parts of body affected, and object/substance that directly injured or made person ill (e.g., Second degree burns on right forearm from acetylene torch)

Classify the case

CHECK ONLY ONE box for each case based on the most serious outcome for that case:

Death (G)	Days away from work (H)	Remained at Work — Job transfer or restriction (I)	Remained at Work — Other recordable cases (J)

Enter the number of days the injured or ill worker was:

Away from work (K)	On job transfer or restriction (L)
days	days

(M) Check the "injury" column or choose one type of illness:

Injury (1)	Skin disorder (2)	Respiratory condition (3)	Poisoning (4)	Hearing loss (5)	All other illnesses (6)

Page totals ▶

Be sure to transfer these totals to the Summary page (Form 300A) before you post it.

Page ____ of ____

Public reporting burden for this collection of information is estimated to average 14 minutes per response, including time to review the instructions, search and gather the data needed, and complete and review the collection of information. Persons are not required to respond to the collection of information unless it displays a currently valid OMB control number. If you have any comments about these estimates or any other aspects of this data collection, contact: US Department of Labor, OSHA Office of Statistical Analysis, Room N-3644, 200 Constitution Avenue, NW, Washington, DC 20210. Do not send the completed forms to this office.

Courtesy of the Occupational Safety & Health Administration, U.S. Department of Labor

U.S. Department of Labor
Occupational Safety and Health Administration

Form approved OMB no. 1218-0176

OSHA's Form 300A (Rev. 01/2004)

Summary of Work-Related Injuries and Illnesses

Year 20____

All establishments covered by Part 1904 must complete this Summary page, even if no work-related injuries or illnesses occurred during the year. Remember to review the Log to verify that the entries are complete and accurate before completing this summary.

Using the Log, count the individual entries you made for each category. Then write the totals below, making sure you've added the entries from every page of the Log. If you had no cases, write "0."

Employees, former employees, and their representatives have the right to review the OSHA Form 300 in its entirety. They also have limited access to the OSHA Form 301 or its equivalent. See 29 CFR Part 1904.35, in OSHA's recordkeeping rule, for further details on the access provisions for these forms.

Number of Cases

Total number of deaths	Total number of cases with days away from work	Total number of cases with job transfer or restriction	Total number of other recordable cases
(G)	(H)	(I)	(J)

Number of Days

Total number of days away from work	Total number of days of job transfer or restriction
(K)	(L)

Injury and Illness Types

Total number of . . .
(M)

(1) Injuries _____
(2) Skin disorders _____
(3) Respiratory conditions _____

(4) Poisonings _____
(5) Hearing loss _____
(6) All other illnesses _____

Post this Summary page from February 1 to April 30 of the year following the year covered by the form.

Public reporting burden for this collection of information is estimated to average 58 minutes per response, including time to review the instructions, search and gather the data needed, and complete and review the collection of information. Persons are not required to respond to the collection of information unless it displays a currently valid OMB control number. If you have any comments about these estimates or any other aspects of this data collection, contact: US Department of Labor, OSHA Office of Statistical Analysis, Room N-3644, 200 Constitution Avenue, NW, Washington, DC 20210. Do not send the completed forms to this office.

Establishment information

Your establishment name _____

Street _____

City _____ State _____ ZIP _____

Industry description (e.g., Manufacture of motor truck trailers) _____

Standard Industrial Classification (SIC), if known (e.g., 3715) _____

OR

North American Industrial Classification (NAICS), if known (e.g., 336212) _____

Employment information (If you don't have these figures, see the Worksheet on the back of this page to estimate.)

Annual average number of employees _____

Total hours worked by all employees last year _____

Sign here

Knowingly falsifying this document may result in a fine.

I certify that I have examined this document and that to the best of my knowledge the entries are true, accurate, and complete.

_____
Company executive   Title

(____)_____   Date
Phone   __/__/__

Optional

Worksheet to Help You Fill Out the Summary

At the end of the year, OSHA requires you to enter the average number of employees and the total hours worked by your employees on the summary. If you don't have these figures, you can use the information on this page to estimate the numbers you will need to enter on the Summary page at the end of the year.

How to figure the average number of employees who worked for your establishment during the year:

❶ Add the total number of employees your establishment paid in all pay periods during the year. Include all employees: full-time, part-time, temporary, seasonal, salaried, and hourly.

The number of employees paid in all pay periods = **❶** _____

❷ Count the number of pay periods your establishment had during the year. Be sure to include any pay periods when you had no employees.

The number of pay periods during the year = **❷** _____

❸ Divide the number of employees by the number of pay periods.

❶ _____ = **❸** _____
❷

❹ Round the answer to the next highest whole number. Write the rounded number in the blank marked *Annual average number of employees.*

The number rounded = **❹** _____

For example, Acme Construction figured its average employment this way:

For pay period...	Acme paid this number of employees...	
1	10	
2	0	
3	15	
4	30	
5	40	
24	▶ 20	
25	15	
26	+10	
	830	

Number of employees paid = 830   **❶**

Number of pay periods = 26   **❷**

830 = 31.92   **❸**
26

31.92 rounds to 32   **❹**

32 is the annual average number of employees

How to figure the total hours worked by all employees:

Include hours worked by salaried, hourly, part-time and seasonal workers, as well as hours worked by other workers subject to day to day supervision by your establishment (e.g., temporary help services workers).

Do not include vacation, sick leave, holidays, or any other non-work time, even if employees were paid for it. If your establishment keeps records of only the hours paid or if you have employees who are not paid by the hour, please estimate the hours that the employees actually worked.

If this number isn't available, you can use this optional worksheet to estimate it.

Optional Worksheet

Find the number of full-time employees in your establishment for the year. _____

Multiply by the number of work hours for a full-time employee in a year. X _____

This is the number of full-time hours worked. _____

Add the number of any overtime hours as well as the hours worked by other employees (part-time, temporary, seasonal) + _____

Round the answer to the next highest whole number. Write the rounded number in the blank marked *Total hours worked by all employees last year.* _____

OSHA's Form 301
Injury and Illness Incident Report

U.S. Department of Labor
Occupational Safety and Health Administration

Form approved OMB no. 1218-0176

Attention: This form contains information relating to employee health and must be used in a manner that protects the confidentiality of employees to the extent possible while the information is being used for occupational safety and health purposes.

This *Injury and Illness Incident Report* is one of the first forms you must fill out when a recordable work-related injury or illness has occurred. Together with the *Log of Work-Related Injuries and Illnesses* and the accompanying *Summary*, these forms help the employer and OSHA develop a picture of the extent and severity of work-related incidents.

Within 7 calendar days after you receive information that a recordable work-related injury or illness has occurred, you must fill out this form or an equivalent. Some state workers' compensation, insurance, or other reports may be acceptable substitutes. To be considered an equivalent form, any substitute must contain all the information asked for on this form.

According to Public Law 91-596 and 29 CFR 1904, OSHA's recordkeeping rule, you must keep this form on file for 5 years following the year to which it pertains.

If you need additional copies of this form, you may photocopy and use as many as you need.

Completed by _____

Title _____

Phone (_____) _____ - _____ Date ____ / ____ / ____

Information about the employee

1) Full name _____

2) Street _____
   City _____ State ____ ZIP ____

3) Date of birth ____ / ____ / ____

4) Date hired ____ / ____ / ____

5) ☐ Male ☐ Female

Information about the physician or other health care professional

6) Name of physician or other health care professional _____

7) If treatment was given away from the worksite, where was it given?
   Facility _____
   Street _____
   City _____ State ____ ZIP ____

8) Was employee treated in an emergency room?
   ☐ Yes ☐ No

9) Was employee hospitalized overnight as an in-patient?
   ☐ Yes ☐ No

Information about the case

10) Case number from the *Log* _____ *(Transfer the case number from the Log after you record the case.)*

11) Date of injury or illness ____ / ____ / ____

12) Time employee began work _____ AM / PM

13) Time of event _____ AM / PM ☐ Check if time cannot be determined

14) **What was the employee doing just before the incident occurred?** Describe the activity, as well as the tools, equipment, or material the employee was using. Be specific. *Examples:* "climbing a ladder while carrying roofing materials"; "spraying chlorine from hand sprayer"; "daily computer key-entry."

15) **What happened?** Tell us how the injury occurred. *Examples:* "When ladder slipped on wet floor, worker fell 20 feet"; "Worker was sprayed with chlorine when gasket broke during replacement"; "Worker developed soreness in wrist over time."

16) **What was the injury or illness?** Tell us the part of the body that was affected and how it was affected; be more specific than "hurt," "pain," or sore." *Examples:* "strained back"; "chemical burn, hand"; "carpal tunnel syndrome."

17) **What object or substance directly harmed the employee?** *Examples:* "concrete floor"; "chlorine"; "radial arm saw." *If this question does not apply to the incident, leave it blank.*

18) **If the employee died, when did death occur?** Date of death ____ / ____ / ____

OSHA JOB SAFETY AND HEALTH POSTER SAMPLES

The **OSHA Job Safety and Health: It's the Law** poster, available for free from OSHA, informs workers of their rights under the Occupational Safety and Health Act. All covered employers are required to display the poster in their workplace. Employers do not need to replace previous versions of the poster. Employers must display the poster in a conspicuous place where workers can see it. Examples are shown here and can be found at dewalt.com/guides; the original version is available from OSHA at http://www.osha.gov.

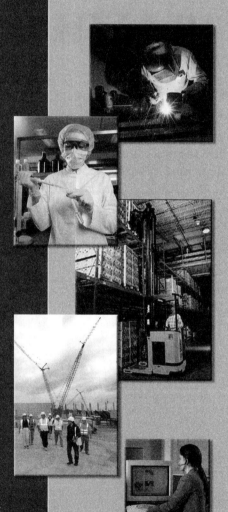

Job Safety and Health
It's the law!

OSHA
Occupational Safety and Health Administration
U.S. Department of Labor

EMPLOYEES:

- You have the right to notify your employer or OSHA about workplace hazards. You may ask OSHA to keep your name confidential.

- You have the right to request an OSHA inspection if you believe that there are unsafe and unhealthful conditions in your workplace. You or your representative may participate in that inspection.

- You can file a complaint with OSHA within 30 days of retaliation or discrimination by your employer for making safety and health complaints or for exercising your rights under the *OSH Act*.

- You have the right to see OSHA citations issued to your employer. Your employer must post the citations at or near the place of the alleged violations.

- Your employer must correct workplace hazards by the date indicated on the citation and must certify that these hazards have been reduced or eliminated.

- You have the right to copies of your medical records and records of your exposures to toxic and harmful substances or conditions.

- Your employer must post this notice in your workplace.

- You must comply with all occupational safety and health standards issued under the *OSH Act* that apply to your own actions and conduct on the job.

EMPLOYERS:

- You must furnish your employees a place of employment free from recognized hazards.

- You must comply with the occupational safety and health standards issued under the *OSH Act*.

This free poster available from OSHA –
The Best Resource for Safety and Health

Free assistance in identifying and correcting hazards or complying with standards is available to employers, without citation or penalty, through OSHA-supported consultation programs in each state.

1-800-321-OSHA
www.osha.gov

OSHA 3165-12-06R

Courtesy of the Occupational Safety & Health Administration, U.S. Department of Labor

Seguridad y Salud en el Trabajo
¡Es la Ley!

OSHA
Administración de Seguridad y Salud Ocupacional
Departamento del Trabajo de los Estados Unidos

EMPLEADOS:

- Usted tiene el derecho de notificar a su empleador o a la OSHA sobre peligros en el lugar de trabajo. Usted también puede pedir que la OSHA no revele su nombre.

- Usted tiene el derecho de pedir a la OSHA que realice una inspección si usted piensa que en su trabajo existen condiciones peligrosas o poco saludables. Usted o su representante pueden participar en esa inspección.

- Usted tiene 30 días para presentar una queja ante la OSHA si su empleador llega a tomar represalias o discriminar en su contra por haber denunciado la condición de seguridad o salud o por ejercer los derechos consagrados bajo la Ley OSH.

- Usted tiene el derecho de ver las citaciones enviadas por la OSHA a su empleador. Su empleador debe colocar las citaciones en el lugar donde se encontraron las supuestas infracciones o cerca del mismo.

- Su empleador debe corregir los peligros en el lugar de trabajo para la fecha indicada en la citación y debe certificar que dichos peligros se hayan reducido o desaparecido.

- Usted tiene derecho de recibir copias de su historial o registro médico y el registro de su exposición a sustancias o condiciones tóxicas o dañinas.

- Su empleador debe colocar este aviso en su lugar de trabajo.

- Usted debe cumplir con todas las normas de seguridad y salud ocupacionales expedidas conforme a la Ley OSH que sean aplicables a sus propias acciones y conducta en el trabajo.

EMPLEADORES:

- Usted debe proporcionar a sus empleados un lugar de empleo libre de peligros conocidos.

- Usted debe cumplir con las normas de seguridad y salud ocupacionales expedidas conforme a la Ley OSH.

Los empleadores pueden obtener ayuda gratis para identificar y corregir las fuentes de peligro y para cumplir con las normas, sin citación ni multa, por medio de programas de consulta respaldados por la OSHA en cada estado del país.

1-800-321-OSHA
www.osha.gov

OSHA 3167-01-07R

HAZARD COMMUNICATION GUIDELINES FOR COMPLIANCE

This checklist can be found at dewalt.com/guides. The full version of OSHA's *Hazard Communication Guidelines for Compliance* can be found at: http://www.osha.gov under Publication 3111.

Checklist for Compliance

The following checklist will help to ensure that you comply with the rule:

- ❏ Obtain a copy of the rule.
- ❏ Read and understand the requirements.
- ❏ Assign responsibility for tasks.
- ❏ Prepare an inventory of chemicals.
- ❏ Ensure containers are labeled.
- ❏ Obtain MSDSs for each chemical.
- ❏ Prepare a written program.
- ❏ Make MSDSs available to workers.
- ❏ Conduct training for workers.
- ❏ Establish procedures to maintain current program.
- ❏ Establish procedures to evaluate effectiveness.

CAL/OSHA MODEL INJURY AND ILLNESS PREVENTION PROGRAM FOR HIGH-HAZARD INDUSTRIES

This model IIPP was produced by Cal/OSHA. This and other model programs can be found at: http://www.dir .ca.gov/DOSH/. This model can also be found at dewalt.com/guides.

MODEL INJURY AND ILLNESS PREVENTION PROGRAM FOR HIGH-HAZARD EMPLOYERS

Responsibility

The Injury and Illness Prevention Program (IIP Program) administrator,

_____

Program Administrator

has the authority and responsibility for implementing the provisions of this program for

_____.

Establishment Name

All managers and supervisors are responsible for implementing and maintaining the IIP Program in their work areas and for answering worker questions about the IIP Program. A copy of this IIP Program is available from each manager and supervisor.

Compliance

Management is responsible for ensuring that all safety and health policies and procedures are clearly communicated and understood by all employees. Managers and supervisors are expected to enforce the rules fairly and uniformly.

All employees are responsible for using safe work practices, for following all directives, policies, and procedures, and for assisting in maintaining a safe work environment.

Our system of ensuring that all workers comply with the rules and maintain a safe work environment include:

- Informing workers of the provisions of our IIP Program;
- Evaluating the safety performance of all workers;

- Recognizing employees who perform safe and healthful work practices;
- Providing training to workers whose safety performance is deficient;
- Disciplining workers for failure to comply with safe and healthful work practices; and
- The following practices: _____

_____

_____

Communication

We recognize that open, two-way communication between management and staff on health and safety issues is essential to an injury-free, productive workplace. The following system of communication is designed to facilitate a continuous flow of safety and health information between management and staff in a form that is readily understandable and consists of one or more of the following checked items:

- New worker orientation including a discussion of safety and health policies and procedures.
- Review of our IIP Program.
- Workplace safety and health training programs.
- Regularly scheduled safety meetings.
- Effective communication of safety and health concerns between workers and supervisors, including translation where appropriate.
- Posted or distributed safety information.
- A system for workers to anonymously inform management about workplace hazards.
- Our establishment has less than ten employees and communicates with and instructs employees orally about general safe work practices and with respect to hazards unique to each employee's job assignment.
- A labor/management safety and health committee that meets regularly, prepares written records of the safety and health committees meetings, reviews results of the periodic scheduled inspections, reviews investigations of accidents and exposures and makes suggestions to management for the prevention of future incidents, reviews investigations of alleged hazardous conditions, and submits recommendations to assist in the evaluation of employee safety suggestions.
- Other: _____

_____

_____

Hazard Assessment

Periodic inspections to identify and evaluate workplace hazards shall be performed by the following competent observer(s) in the following areas of our workplace:

Competent Observer	Area

Periodic inspections are performed according to the following schedule:

- _____;
  Frequency (daily, weekly, monthly, etc.)
- When we initially established our IIP Program;
- When new substances, processes, procedures, or equipment that present potential new hazards are introduced into our workplace;
- When new, previously unidentified hazards are recognized;
- When occupational injuries and illnesses occur;
- When we hire and/or reassign permanent or intermittent workers to processes, operations, or tasks for which a hazard evaluation has not been previously conducted; and
- Whenever workplace conditions warrant an inspection.

Periodic inspections consist of identification and evaluation of workplace hazards utilizing applicable sections of the attached Hazard Assessment Checklist and any other effective methods to identify and evaluate workplace hazards.

Accident/Exposure Investigations

Procedures for investigating workplace accidents and hazardous substance exposures include:

- Visiting the accident scene as soon as possible;
- Interviewing injured workers and witnesses;
- Examining the workplace for factors associated with the accident/exposure;
- Determining the cause of the accident/exposure;
- Taking corrective action to prevent the accident/exposure from reoccurring; and
- Recording the findings and corrective actions taken.

Hazard Correction

Unsafe or unhealthy work conditions, practices, or procedures shall be corrected in a timely manner based on the severity of the hazards. Hazards shall be corrected according to the following procedures:

- When observed or discovered;
- When an imminent hazard exists which cannot be immediately abated without endangering employee(s) and/or property, we will remove all exposed workers from the area except those necessary to correct the existing condition. Workers necessary to correct the hazardous condition shall be provided with the necessary protection; and
- All such actions taken and dates they are completed shall be documented on the appropriate forms.

Training and Instruction

All workers, including managers and supervisors, shall have training and instruction on general and job-specific safety and health practices. Training and instruction shall be provided as follows:

- When the IIP Program is first established;
- To all new workers, except for construction workers who are provided training through a Cal/OSHA-approved construction industry occupational safety and health training program;

- To all workers given new job assignments for which training has not been previously provided;
- Whenever new substances, processes, procedures, or equipment are introduced to the workplace and represent a new hazard;
- Whenever the employer is made aware of a new or previously unrecognized hazard;
- To supervisors to familiarize them with the safety and health hazards to which workers under their immediate direction and control may be exposed; and
- To all workers with respect to hazards specific to each employee's job assignment.

Workplace safety and health practices for all industries include, but are not limited to, the following:

- Explanation of the employer's IIP Program, emergency action plan, and fire prevention plan and measures for reporting any unsafe conditions, work practices, injuries, and when additional instruction is needed.
- Use of appropriate clothing, including gloves, footwear, and personal protective equipment.
- Information about chemical hazards to which employees could be exposed and other hazard communication program information.
- Availability of toilet, handwashing, and drinking water facilities.
- Provisions for medical services and first aid including emergency procedures.

In addition, we provide specific instructions to all workers regarding hazards unique to their job assignment, to the extent that such information was not already covered in other training.

Recordkeeping

We have checked one of the following categories as our recordkeeping policy.

- Category 1. Our establishment is on a designated high-hazard industry list. We have taken the following steps to implement and maintain our IIP Program:

    1. Records of hazard assessment inspections, including the person or persons conducting the inspection, the unsafe conditions and work practices that have been identified, and the action taken to correct the identified unsafe conditions and work practices, are recorded on a hazard assessment and correction form; and

    2. Documentation of safety and health training for each worker, including the worker's name or other identifier, training dates, type(s) of training, and training providers, are recorded on a worker training and instruction form. We also include the records relating to worker training provided by a construction industry occupational safety and health program approved by Cal/OSHA.

    Inspection records and training documentation will be maintained according to the following checked schedule:

    ○ For one year, except for training records of employees who have worked for less than one year which are provided to the worker upon termination of employment; or

    ○ Since we have less than ten workers, including managers and supervisors, we maintain inspection records only until the hazard is corrected and only maintain a log of instructions to workers with respect to worker job assignments when they are first hired or assigned new duties.

- Category 2. We are a local governmental entity (any county, city, or district, and any public or quasi-public corporation or public agency therein) and we are not required to keep written records of the steps taken to implement and maintain our IIP Program.

List of Training Subjects

We train our workers about the following checked training subjects:

- The employer's Code of Safe Practices.
- Confined spaces.
- Safe practices for operating any agricultural equipment.
- Good housekeeping, fire prevention, safe practices for operating any construction equipment.
- Safe procedures for cleaning, repairing, servicing, and adjusting equipment and machinery.
- Safe access to working areas.
- Protection from falls.
- Electrical hazards, including working around high-voltage lines.
- Crane operations.
- Trenching and excavation work.
- Proper use of powered tools.
- Guarding of belts and pulleys, gears and sprockets, and conveyor nip points.
- Guarding of machine, machine parts, and prime movers.
- Lock-out/tag-out procedures.
- Materials handling.
- Chainsaw and other power tool operation.
- Tree falling/bucking procedures and precautions, including procedures for recognizing and working with hazard trees, snags, lodged trees, and unsafe weather conditions.
- Yarding operations, including skidding, running lines, unstable logs, rigging, and communication.
- Landing and loading areas, including release of rigging, landing layout, moving vehicles and equipment, and log truck locating, loading, and wrapping.
- Fall protection from elevated locations.
- Use of elevated platforms, including condors and scissor lifts.
- Safe use of explosives.
- Driver safety.
- Slips, falls, and back injuries.
- Ergonomic hazards, including proper lifting techniques and working on ladders or in a stooped posture for prolonged periods at one time.
- Personal protective equipment.
- Respiratory equipment.
- Hazardous chemical exposures.
- Hazard communication.
- Physical hazards, such as heat/cold stress, noise, and ionizing and nonionizing radiation.
- Laboratory safety.
- Bloodborne pathogens and other biological hazards.
- Other job-specific hazards, such as _____

_____

_____

Hazard Assessment Checklist

General Work Environment

- Are all worksites clean and orderly?
- Are work surfaces kept dry or appropriate means taken to ensure the surfaces are slip-resistant?
- Are all spilled materials or liquids cleaned up immediately?
- Is combustible scrap, debris, and waste stored safely and removed from the worksite promptly?
- Is accumulated combustible dust routinely removed from elevated surfaces, including the overhead structure of buildings?
- Is combustible dust cleaned up with a vacuum system to prevent the dust going into suspension?
- Is metallic or conductive dust prevented from entering or accumulation on or around electrical enclosures or equipment?
- Are covered metal waste cans used for oily and paint-soaked waste?
- Are all oil- and gas-fired devices equipped with flame failure controls that will prevent flow of fuel if pilots or main burners are not working?
- Are paint spray booths, dip tanks, and the like cleaned regularly?
- Are the minimum number of toilets and washing facilities provided?
- Are all toilets and washing facilities clean and sanitary?
- Are all work areas adequately illuminated?
- Are pits and floor openings covered or otherwise guarded?

Personal Protective Equipment and Clothing

- Are protective goggles or face shields provided and worn where there is any danger of flying particles or corrosive materials?
- Are approved safety glasses required to be worn at all times in areas where there is a risk of eye injuries such as punctures, abrasions, contusions, or burns?
- Are employees who need corrective lenses (glasses or contacts lenses) in working environments with harmful exposures required to wear only approved safety glasses, protective goggles, or other medically approved precautionary procedures?
- Are protective gloves, aprons, shields, or other means provided against cuts, corrosive liquids, and chemicals?
- Are hard hats provided and worn where danger of falling objects exists?
- Are hard hats inspected periodically for damage to the shell and suspension system?
- Is appropriate foot protection required where there is the risk of foot injuries from hot, corrosive, poisonous substances, falling objects, or crushing or penetrating actions?
- Are approved respirators provided for regular or emergency use where needed?
- Is all protective equipment maintained in a sanitary condition and ready for use?
- Do you have eyewash facilities and a quick-drench shower within the work area where employees are exposed to injurious corrosive materials?
- Where special equipment is needed for electrical workers, is it available?

- When lunches are eaten on the premises, are they eaten in areas where there is no exposure to toxic materials or other health hazards?
- Is protection against the effects of occupational noise exposure provided when sound levels exceed those of the Cal/OSHA noise standard?

Walkways

- Are aisles and passageways kept clear?
- Are aisles and walkways marked as appropriate?
- Are wet surfaces covered with nonslip materials?
- Are holes in the floor, sidewalk, or other walking surface repaired properly, covered, or otherwise made safe?
- Is there safe clearance for walking in aisles where motorized or mechanical handling equipment is operating?
- Are spilled materials cleaned up immediately?
- Are materials or equipment stored in such a way that sharp projectiles will not interfere with the walkway?
- Are changes of direction or elevations readily identifiable?
- Are aisles or walkways that pass near moving or operating machinery, welding operations, or similar operations arranged so employees will not be subjected to potential hazards?
- Is adequate headroom provided for the entire length of any aisle or walkway?
- Are standard guardrails provided wherever aisle or walkway surfaces are elevated more than 30 inches above any adjacent floor or the ground?
- Are bridges provided over conveyors and similar hazards?

Floor and Wall Openings

- Are floor openings guarded by a cover, guardrail, or equivalent on all sides (except at entrance to stairways or ladders)?
- Are toeboards installed around the edges of a permanent floor opening (where persons may pass below the opening)?
- Are skylight screens of such construction and mounting that they will withstand a load of at least 200 pounds?
- Is the glass in windows, doors, and glass walls that are subject to human impact of sufficient thickness and type for the condition of use?
- Are grates or similar type covers over floor openings such as floor drains of such design that foot traffic or rolling equipment will not be affected by the grate spacing?
- Are unused portions of service pits and pits not actually in use either covered or protected by guardrails or equivalent?
- Are manhole covers, trench covers, and similar covers, plus their supports, designed to carry a truck rear axle load of at least 20,000 pounds when located in roadways and subject to vehicle traffic?
- Are floor or wall openings in fire-resistive construction provided with doors or covers compatible with the fire rating of the structure and provided with a self-closing feature when appropriate?

Stairs and Stairways

- Are standard stair rails or handrails on all stairways having four or more risers?
- Are all stairways at least 22" wide?
- Do stairs have at least a 6'6" overhead clearance?
- Do stairs angle no more than 50 and no less than 30 degrees?
- Are stairs of hollow-pan type treads and landings filled to noising level with solid material?
- Are step risers on stairs uniform from top to bottom, with no riser spacing greater than 7½"?
- Are steps on stairs and stairways designed or provided with a surface that renders them slip-resistant?
- Are stairway handrails located between 30 and 34" above the leading edge of stair treads?
- Do stairway handrails have a least 1½" of clearance between the handrails and the wall or surface they are mounted on?
- Are stairway handrails capable of withstanding a load of 200 pounds, applied in any direction?
- Where stairs or stairways exit directly into any area where vehicles may be operated, are adequate barriers and warnings provided to prevent employees stepping into the path of traffic?
- Do stairway landings have a dimension measured in the direction of travel at least equal to width of the stairway?
- Is the vertical distance between stairway landings limited to 12' or less?

Elevated Surfaces

- Are signs posted, when appropriate, showing the elevated surface load capacity?
- Are surfaces elevated more than 30" above the floor or ground provided with standard guardrails?
- Are all elevated surfaces (beneath which people or machinery could be exposed to falling objects) provided with standard 4" toeboards?
- Is a permanent means of access and egress provided to elevated storage and work surfaces?
- Is required headroom provided where necessary?
- Is material on elevated surfaces piled, stacked, or racked in a manner to prevent it from tipping, falling, collapsing, rolling, or spreading?
- Are dock boards or bridge plates used when transferring materials between docks and trucks or rail cars?

Exiting or Egress

- Are all exits marked with an exit sign and illuminated by a reliable light source?
- Are the directions to exits, when not immediately apparent, marked with visible signs?
- Are doors, passageways, or stairways that are neither exits nor access to exits and which could be mistaken for exits appropriately marked "NOT AN EXIT," "TO BASEMENT," "STOREROOM," and the like?
- Are exit signs provided with the word "EXIT" in lettering at least 5" high and the stroke of the lettering at least ½" wide?
- Are exit doors side-hinged?
- Are all exits kept free of obstructions?
- Are at least two means of egress provided from elevated platforms, pits, or rooms where the absence of a second exit would increase the risk of injury from hot, poisonous, corrosive, suffocating, flammable, or explosive substances?
- Are there sufficient exits to permit prompt escape in case of emergency?
- Are special precautions taken to protect employees during construction and repair operations?

- Is the number of exits from each floor of a building, and the number of exits from the building itself, appropriate for the building occupancy load?

- Are exit stairways that are required to be separated from other parts of a building enclosed by at least 2-hour fire-resistive construction in buildings more than four stories in height and not less than 1-hour fire-resistive construction elsewhere?

- When ramps are used as part of required exiting from a building, is the ramp slope limited to 1' vertical and 12' horizontal?

- Where exiting will be through frameless glass doors, glass exit doors, storm doors, and such, are the doors fully tempered and do they meet the safety requirements for human impact?

Exit Doors

- Are doors that are required to serve as exits designed and constructed so that the way of exit travel is obvious and direct?

- Are windows that could be mistaken for exit doors made inaccessible by means of barriers or railings?

- Are exit doors openable from the direction of exit travel without the use of a key or any special knowledge or effort when the building is occupied?

- Is a revolving, sliding, or overhead door prohibited from serving as a required exit door?

- Where panic hardware is installed on a required exit door, will it allow the door to open by applying a force of 15 pounds or less in the direction of the exit traffic?

- Are doors on cold storage rooms provided with an inside release mechanism that will release the latch and open the door even if it's padlocked or otherwise locked on the outside?

- Where exit doors open directly onto any street, alley, or other area where vehicles may be operated, are adequate barriers and warnings provided to prevent employees stepping into the path of traffic?

- Are doors that swing in both directions and are located between rooms where there is frequent traffic provided with viewing panels in each door?

Portable Ladders

- Are all ladders maintained in good condition, joints between steps and side rails tight, all hardware and fittings securely attached, and moveable parts operating freely without binding or undue play?

- Are nonslip safety feet provided on each ladder?

- Are nonslip safety feet provided on each metal or rung ladder?

- Are ladder rungs and steps free of grease and oil?

- Is it prohibited to place a ladder in front of doors opening toward the ladder except when the door is blocked open, locked, or guarded?

- Is it prohibited to place ladders on boxes, barrels, or other unstable bases to obtain additional height?

- Are employees instructed to face the ladder when ascending or descending?

- Are employees prohibited from using ladders that are broken; have missing steps, rungs, or cleats; or have broken side rails or other faulty equipment?

- Are employees instructed not to use the top two steps of ordinary stepladders as a step?

- When portable rung ladders are used to gain access to elevated platforms, roofs, and the like, does the ladder always extend at least 3' above the elevated surface?

- Is it required that when portable rung or cleat-type ladders are used the base is so placed that slipping will not occur or it is lashed or otherwise held in place?

- Are portable metal ladders legibly marked with signs reading "CAUTION Do Not Use Around Electrical Equipment" or equivalent wording?
- Are employees prohibited from using ladders as guys, braces, skids, gin poles, or for other than their intended purposes?
- Are employees instructed to only adjust extension ladders while standing at a base (not while standing on the ladder or from a position above the ladder)?
- Are metal ladders inspected for damage?
- Are the rungs of ladders uniformly spaced at 12", center to center?

Hand Tools and Equipment

- Are all tools and equipment (both company- and employee-owned) used by employees at their workplace in good condition?
- Are hand tools such as chisels and punches that develop mushroomed heads during use reconditioned or replaced as necessary?
- Are broken or fractured handles on hammers, axes, and similar equipment replaced promptly?
- Are worn or bent wrenches replaced regularly?
- Are appropriate handles used on files and similar tools?
- Are employees made aware of the hazards caused by faulty or improperly used hand tools?
- Are appropriate safety glasses, face shields, and similar equipment used while using hand tools or equipment that might produce flying materials or be subject to breakage?
- Are jacks checked periodically to ensure they are in good operating condition?
- Are tool handles wedged tightly in the head of all tools?
- Are tool cutting edges kept sharp so the tool will move smoothly without binding or skipping?
- Are tools stored in a dry, secure location where they won't be tampered with?
- Is eye and face protection used when driving hardened or tempered spuds or nails?

Portable (Power Operated) Tools and Equipment

- Are grinders, saws, and similar equipment provided with appropriate safety guards?
- Are power tools used with the correct shield, guard, or attachment recommended by the manufacturer?
- Are portable circular saws equipped with guards above and below the base shoe?
- Are circular saw guards checked to ensure they are not wedged up, thus leaving the lower portion of the blade unguarded?
- Are rotating or moving parts of equipment guarded to prevent physical contact?
- Are all cord-connected, electrically operated tools and equipment effectively grounded or of the approved double-insulated type?
- Are effective guards in place over belts, pulleys, chains, and sprockets on equipment such as concrete mixers, air compressors, and the like?
- Are portable fans provided with full guards or screens having openings ½" or less?
- Is hoisting equipment available and used for lifting heavy objects, and are hoist ratings and characteristics appropriate for the task?
- Are ground-fault circuit interrupters provided on all temporary electrical 15- and 20-ampere circuits used during periods of construction?

- Are pneumatic and hydraulic hoses on power-operated tools checked regularly for deterioration or damage?

Abrasive Wheel Equipment Grinders

- Is the work rest used and kept adjusted to within $1/8$" of the wheel?
- Is the adjustable tongue on the top side of the grinder used and kept adjusted to within ¼" of the wheel?
- Do side guards cover the spindle, nut, and flange and 75% of the wheel diameter?
- Are bench and pedestal grinders permanently mounted?
- Are goggles or face shields always worn when grinding?
- Is the maximum RPM rating of each abrasive wheel compatible with the RPM rating of the grinder motor?
- Are fixed or permanently mounted grinders connected to their electrical supply system with metallic conduit or other permanent wiring method?
- Does each grinder have an individual on and off control switch?
- Is each electrically operated grinder effectively grounded?
- Before new abrasive wheels are mounted, are they visually inspected and ring tested?
- Are dust collectors and powered exhausts provided on grinders used in operations that produce large amounts of dust?
- Are splashguards mounted on grinders that use coolant to prevent the coolant reaching employees?
- Is cleanliness maintained around grinder?

Powder-Actuated Tools

- Are employees who operate powder-actuated tools trained in their use and carry a valid operator's card?
- Do the powder-actuated tools being used have written approval of the Division of Occupational Safety and Health?
- Is each powder-actuated tool stored in its own locked container when not being used?
- Is a sign at least 7" by 10" with bold type reading "POWDER-ACTUATED TOOL IN USE" conspicuously posted when the tool is being used?
- Are powder-actuated tools left unloaded until they are actually ready to be used?
- Are powder-actuated tools inspected for obstructions or defects each day before use?
- Do powder-actuated tool operators have and use appropriate personal protective equipment such as hard hats, safety goggles, safety shoes, and ear protectors?

Machine Guarding

- Is there a training program to instruct employees on safe methods of machine operation?
- Is there adequate supervision to ensure that employees are following safe machine operating procedures?
- Is there a regular program of safety inspection of machinery and equipment?
- Is all machinery and equipment kept clean and properly maintained?
- Is sufficient clearance provided around and between machines to allow for safe operations, setup and servicing, material handling, and waste removal?
- Is equipment and machinery securely placed and anchored when necessary to prevent tipping or other movement that could result in personal injury?
- Is there a power shutoff switch within reach of the operator's position at each machine?

- Can electric power to each machine be locked out for maintenance, repair, or security?
- Are the non-current-carrying metal parts of electrically operated machines bonded and grounded?
- Are foot-operated switches guarded or arranged to prevent accidental actuation by personnel or falling objects?
- Are manually operated valves and switches controlling the operation of equipment and machines clearly identified and readily accessible?
- Are all emergency stop buttons colored red?
- Are all pulleys and belts that are within 7' of the floor or working level properly guarded?
- Are all moving chains and gears properly guarded?
- Are splashguards mounted on machines that use coolant to prevent the coolant from reaching employees?
- Are methods provided to protect the operator and other employees in the machine area from hazards created at the point of operation, ingoing nip points, rotating parts, flying chips, and sparks?
- Are machinery guards secure and so arranged that they do not offer a hazard in their use?
- If special hand tools are used for placing and removing material, do they protect the operator's hands?
- Are revolving drums, barrels, and containers required to be guarded by an enclosure that is interlocked with the drive mechanism so that revolution cannot occur unless the guard enclosure is in place so guarded?
- Do arbors and mandrels have firm and secure bearings and are they free from play?
- Are provisions made to prevent machines from automatically starting when power is restored after a power failure or shutdown?
- Are machines constructed so as to be free from excessive vibration when the largest-size tool is mounted and run at full speed?
- If machinery is cleaned with compressed air, is air pressure controlled and personal protective equipment or other safeguards used to protect operators and other workers from eye and body injury?
- Are fan blades protected with a guard having openings no larger than ½" when operating within 7' of the floor?
- Are saws used for ripping equipped with anti-kickback devices and spreaders?
- Are radial arm saws so arranged that the cutting head will gently return to the back of the table when released?

Lockout Blockout Procedures

- Is all machinery or equipment capable of movement required to be de-energized or disengaged and blocked or locked out during cleaning, servicing, adjusting, or setting-up operations whenever required?
- Is the locking out of control circuits in lieu of locking out main power disconnects prohibited?
- Are all equipment control valve handles provided with a means for locking out?
- Does the lockout procedure require that stored energy (i.e., mechanical, hydraulic, air) be released or blocked before equipment is locked out for repairs?
- Are appropriate employees provided with individually keyed personal safety locks?
- Are employees required to keep personal control of their key(s) while they have safety locks in use?
- Is it required that employees check the safety of the lockout by attempting a start-up after making sure no one is exposed?
- Where the power disconnecting means for equipment does not also disconnect the electrical control circuit:

- ○ Are the appropriate electrical enclosures identified?
- ○ Are means provided to ensure the control circuit can also be disconnected and locked out?

Welding, Cutting, and Brazing

- Are only authorized and trained personnel permitted to use welding, cutting, or brazing equipment?
- Do all operators have a copy of the appropriate operating instructions and are they directed to follow them?
- Are compressed gas cylinders regularly examined for obvious signs of defects, deep rusting, or leakage?
- Is care used in handling and storage of cylinders, safety valves, relief valves, and the like to prevent damage?
- Are precautions taken to prevent the mixture of air or oxygen with flammable gases except at a burner or in a standard torch?
- Is only approved apparatus (torches, regulators, pressure-reducing valves, acetylene generators, manifolds) used?
- Are cylinders kept away from sources of heat?
- Is it prohibited to use cylinders as rollers or supports?
- Are empty cylinders appropriately marked, their valves closed, and valve-protection caps on?
- Are signs reading: "DANGER NO-SMOKING, MATCHES, OR OPEN LIGHTS" or the equivalent posted?
- Are cylinders, cylinder valves, couplings, regulators, hoses, and apparatus kept free of oily or greasy substances?
- Is care taken not to drop or strike cylinders?
- Unless secured on special trucks, are regulators removed and valve-protection caps put in place before moving cylinders?
- Do cylinders without fixed hand wheels have keys, handles, or non-adjustable wrenches on stem valves when in service?
- Are liquefied gases stored and shipped valve-end-up with valve covers in place?
- Are employees instructed to never crack a fuel-gas cylinder valve near sources of ignition?
- Before a regulator is removed, is the valve closed and gas released from the regulator?
- Is red used to identify the acetylene (and other fuel-gas) hose, green for oxygen hose, and black for inert gas and air hose?
- Are pressure-reducing regulators used only for the gas and pressures for which they are intended?
- Is the open-circuit (no-load) voltage of arc welding and cutting machines as low as possible and not in excess of the recommended limits?
- Under wet conditions, are automatic controls for reducing no-load voltage used?
- Are grounding of the machine frame and safety ground connections of portable machines checked periodically?
- Are electrodes removed from the holders when not in use?
- Is it required that electric power to the welder be shut off when no one is in attendance?
- Is suitable fire extinguishing equipment available for immediate use?
- Is the welder forbidden to coil or loop welding electrode cable around his body?
- Are wet machines thoroughly dried and tested before being used?
- Are work and electrode lead cables frequently inspected for wear and damage and replaced when needed?

- Do means for connecting cable lengths have adequate insulation?
- When the object to be welded cannot be moved and fire hazards cannot be removed, are shields used to confine heat, sparks, and slag?
- Are firewatchers assigned when welding or cutting is performed in locations where a serious fire might develop?
- Are combustible floors kept wet, covered by damp sand, or protected by fire-resistant shields?
- When floors are wet down, are personnel protected from possible electrical shock?
- When welding is done on metal walls, are precautions taken to protect combustibles on the other side?
- Before hot work is begun, are used drums, barrels, tanks, and other containers so thoroughly cleaned that no substances remain that could explode, ignite, or produce toxic vapors?
- Is it required that eye protection helmets, hand shields, and goggles meet appropriate standards?
- Are employees exposed to the hazards created by welding, cutting, or bracing operations protected with personal protective equipment and clothing?
- Is a check made for adequate ventilation in and where welding or cutting is performed?
- When working in confined places, are environmental monitoring tests taken and means provided for quick removal of welders in case of an emergency?

Compressors and Compressed Air

- Are compressors equipped with pressure-relief valves and pressure gauges?
- Are compressor air intakes installed and equipped to ensure that only clean, uncontaminated air enters the compressor?
- Are air filters installed on the compressor intake?
- Are compressors operated and lubricated in accordance with the manufacturer's recommendations?
- Are safety devices on compressed air systems checked frequently?
- Before any repair work is done on the pressure system of a compressor, is the pressure bled off and the system locked out?
- Are signs posted to warn of the automatic starting feature of compressors?
- Is the belt drive system totally enclosed to provide protection for the front, back, top, and sides?
- Is it strictly prohibited to direct compressed air toward a person?
- Are employees prohibited from using highly compressed air for cleaning purposes?
- If compressed air is used for cleaning off clothing, is the pressure reduced to less than 10 psi?
- When using compressed air for cleaning, do employees use personal protective equipment?
- Are safety chains or other suitable locking devices used at couplings of high-pressure hose lines where a connection failure would create a hazard?
- Before compressed air is used to empty containers of liquid, is the safe working pressure of the container checked?
- When compressed air is used with abrasive blast cleaning equipment, is the operating valve a type that must be held open manually?
- When compressed air is used to inflate auto tires, is a clip-on chuck and an inline regulator preset to 40 psi required?
- Is it prohibited to use compressed air to clean up or move combustible dust if such action could cause the dust to be suspended in the air and cause a fire or explosion hazard?

Compressed Air Receivers

- Is every receiver equipped with a pressure gauge and with one or more automatic, spring-loaded safety valves?

- Is the total relieving capacity of the safety valve capable of preventing pressure in the receiver from exceeding the maximum allowable working pressure of the receiver by more than 10%?

- Is every air receiver provided with a drainpipe and valve at the lowest point for the removal of accumulated oil and water?

- Are compressed air receivers periodically drained of moisture and oil?

- Are all safety valves tested frequently and at regular intervals to determine whether they are in good operating condition?

- Is there a current operating permit issued by the Division of Occupational Safety and Health?

- Is the inlet of air receivers and piping systems kept free of accumulated oil and carbonaceous materials?

Compressed Gas and Cylinders

- Are cylinders with a water weight capacity over 30 pounds equipped with means for connecting a valve protector device or with a collar or recess to protect the valve?

- Are cylinders legibly marked to clearly identify the gas contained?

- Are compressed gas cylinders stored in areas that are protected from external heat sources such as flame impingement, intense radiant heat, electric arcs, or high-temperature lines?

- Are cylinders located or stored in areas where they will not be damaged by passing or falling objects or subject to tampering by unauthorized persons?

- Are cylinders stored or transported in a manner to prevent them from creating a hazard by tipping, falling, or rolling?

- Are cylinders containing liquefied fuel gas stored or transported in a position so that the safety-relief device is always in direct contact with the vapor space in the cylinder?

- Are valve protectors always placed on cylinders when the cylinders are not in use or connected for use?

- Are all valves closed off before a cylinder is moved, when the cylinder is empty, and at the completion of each job?

- Are low-pressure fuel-gas cylinders checked periodically for corrosion, general distortion, cracks, or any other defect that might indicate a weakness or render it unfit for service?

- Does the periodic check of low-pressure fuel-gas cylinders include a close inspection of the cylinders' bottom?

Hoist and Auxiliary Equipment

- Is each overhead electric hoist equipped with a limit device to stop the hook travel at its highest and lowest point of safe travel?

- Will each hoist automatically stop and hold any load up to 125% of its rated load if its actuating force is removed?

- Is the rated load of each hoist legibly marked and visible to the operator?

- Are stops provided at the safe limits of travel for a trolley hoist?

- Are the controls of hoists plainly marked to indicate the direction of travel or motion?

- Is each cage-controlled hoist equipped with an effective warning device?

- Are close-fitting guards or other suitable devices installed on hoists to ensure hoist ropes will be maintained in the sheave groves?

- Are all hoist chains or ropes of sufficient length to handle the full range of movement for the application while still maintaining two full wraps on the drum at all times?
- Are nip points or contact points between hoist ropes and sheaves that are permanently located within 7' of the floor, ground, or working platform guarded?
- Is it prohibited to use chains or rope slings that are kinked or twisted?
- Is it prohibited to use the hoist rope or chain wrapped around the load as a substitute for a sling?
- Is the operator instructed to avoid carrying loads over people?
- Are only employees who have been trained in the proper use of hoists allowed to operate them?

Industrial Trucks—Forklifts

- Are only trained personnel allowed to operate industrial trucks?
- Is substantial overhead protective equipment provided on high-lift rider equipment?
- Are the required lift truck operating rules posted and enforced?
- Is directional lighting provided on each industrial truck that operates in an area with less than 2 foot-candles per square foot of general lighting?
- Does each industrial truck have a warning horn, whistle, gong, or other device that can be clearly heard above the normal noise in the areas where operated?
- Are the brakes on each industrial truck capable of bringing the vehicle to a complete and safe stop when fully loaded?
- Will the industrial truck's parking brake effectively prevent the vehicle from moving when unattended?
- Are industrial trucks operating in areas where flammable gases or vapors or combustible dust or ignitable fibers may be present in the atmosphere approved for such locations?
- Are motorized hand and hand/rider trucks so designed that the brakes are applied and power to the drive motor shuts off when the operator releases his/her grip on the device that controls the travel?
- Are industrial trucks with internal combustion engines operated in buildings or enclosed areas carefully checked to ensure such operations do not cause harmful concentration of dangerous gases or fumes?

Spraying Operations

- Is adequate ventilation assured before spray operations are started?
- Is mechanical ventilation provided when a spraying operation is done in enclosed areas?
- When mechanical ventilation is provided during spraying operations, is it so arranged that it will not circulate the contaminated air?
- Is the spray area free of hot surfaces?
- Is the spray area at least 20' from flames, sparks, operating electrical motors, and other ignition sources?
- Are portable lamps used to illuminate spray areas suitable for use in a hazardous location?
- Is approved respiratory equipment provided and used when appropriate during spraying operations?
- Do solvents used for cleaning have a flash point of 100°F or more?
- Are fire control sprinkler heads kept clean?
- Are "NO SMOKING" signs posted in spray areas, paint rooms, paint booths, and paint storage areas?
- Is the spray area kept clean of combustible residue?
- Are spray booths constructed of metal, masonry, or other substantial noncombustible material?
- Are spray booth floors and baffles noncombustible and easily cleaned?

- Is infrared drying apparatus kept out of the spray area during spraying operations?
- Is the spray booth completely ventilated before using the drying apparatus?
- Is the electric drying apparatus properly grounded?
- Are lighting fixtures for spray booths located outside of the booth and the interior lighted through sealed clear panels?
- Are the electric motors for exhaust fans placed outside booths or ducts?
- Are belts and pulleys inside the booth fully enclosed?
- Do ducts have access doors to allow cleaning?
- Do all drying spaces have adequate ventilation?

Entering Confined Spaces

- Are confined spaces thoroughly emptied of any corrosive or hazardous substances, such as acids or caustics, before entry?
- Before entry, are all lines to a confined space containing inert, toxic, flammable, or corrosive materials valved off and blanked or disconnected and separated?
- Is it required that all impellers, agitators, or other moving equipment inside confined spaces be locked out if they present a hazard?
- Is either natural or mechanical ventilation provided prior to confined-space entry?
- Before entry, are appropriate atmospheric tests performed to check for oxygen deficiency, toxic substances, and explosive concentrations in the confined space?
- Is adequate illumination provided for the work to be performed in the confined space?
- Is the atmosphere inside the confined space frequently tested or continuously monitored during conduct of the work?
- Is there an assigned safety standby employee outside of the confined space whose sole responsibility is to watch the work in progress, sound an alarm if necessary, and render assistance?
- Are the standby employee or other employees prohibited from entering the confined space without lifelines and respiratory equipment if there is any questions as to the cause of an emergency?
- In addition to the standby employee, is there at least one other trained rescuer in the vicinity?
- Are all rescuers appropriately trained and using approved, recently inspected equipment?
- Does all rescue equipment allow for lifting employees vertically from a top opening?
- Are personnel trained in first aid and CPR immediately available?
- Is there an effective communication system in place whenever respiratory equipment is used and the employee in the confined space is out of sight of the standby person?
- Is approved respiratory equipment required if the atmosphere inside the confined space cannot be made acceptable?
- Is all portable electrical equipment used inside confined spaces either grounded and insulated or equipped with groundfault protection?
- Before gas welding or burning is started in a confined space, are hoses checked for leaks, compressed gas bottles forbidden inside of the confined space, torches lighted only outside of the confined area, and the confined area tested for an explosive atmosphere each time before a lighted torch is to be taken into the confined space?
- If employees will be using oxygen-consuming equipment such as salamanders, torches, or furnaces in a confined space, is sufficient air provided to ensure combustion without reducing the oxygen concentration of the atmosphere below 19.5% by volume?

- Whenever combustion-type equipment is used in a confined space, are provisions made to ensure the exhaust gases are vented outside of the enclosure?
- Is each confined space checked for decaying vegetation or animal matter, which may produce methane?
- Is the confined space checked for possible industrial waste, which could contain toxic properties?
- If the confined space is below the ground and near areas where motor vehicles will be operating, is it possible for vehicle exhaust or carbon monoxide to enter the space?

Environmental Controls

- Are all work areas properly illuminated?
- Are employees instructed in proper first aid and other emergency procedures?
- Are hazardous substances identified that may cause harm by inhalation, ingestion, skin absorption, or contact?
- Are employees aware of the hazards involved with the various chemicals they may be exposed to in their work environment, such as ammonia, chlorine, epoxys, and caustics?
- Is employee exposure to chemicals in the workplace kept within acceptable levels?
- Can a less harmful method or product be used?
- Is the work area's ventilation system appropriate for the work being performed?
- Are spray painting operations done in spray rooms or booths equipped with an appropriate exhaust system?
- Is employee exposure to welding fumes controlled by ventilation, use of respirators, exposure time, or other means?
- Are welders and other workers nearby provided with flash shields during welding operations?
- If forklifts and other vehicles are used in buildings or other enclosed areas, are the carbon monoxide levels kept below maximum acceptable concentrations?
- Has there been a determination that noise levels in the facilities are within acceptable levels?
- Are steps being taken to use engineering controls to reduce excessive noise levels?
- Are proper precautions being taken when handling asbestos and other fibrous materials?
- Are caution labels and signs used to warn of asbestos?
- Are wet methods used, when practicable, to prevent the emission of airborne asbestos fibers, silica dust, and similar hazardous materials?
- Is vacuuming with appropriate equipment used whenever possible rather than blowing or sweeping dust?
- Are grinders, saws, and other machines that produce respirable dusts vented to an industrial collector or central exhaust system?
- Are all local exhaust ventilation systems designed and operating properly, such as airflow and volume necessary for the application? Are the ducts free of obstructions or the belts slipping?
- Is personal protective equipment provided, used, and maintained wherever required?
- Are there written standard operating procedures for the selection and use of respirators where needed?
- Are restrooms and washrooms kept clean and sanitary?
- Is all water provided for drinking, washing, and cooking potable?
- Are all outlets for water not suitable for drinking clearly identified?
- Are employees' physical capacities assessed before being assigned to jobs requiring heavy work?
- Are employees instructed in the proper manner of lifting heavy objects?

- Where heat is a problem, have all fixed work areas been provided with spot cooling or air-conditioning?
- Are employees screened before assignment to areas of high heat to determine if their health condition might make them more susceptible to having an adverse reaction?
- Are employees working on streets and roadways where they are exposed to the hazards of traffic required to wear a bright-colored (traffic orange) warning vest?
- Are exhaust stacks and air intakes located so that contaminated air will not be recirculated within a building or other enclosed area?
- Is equipment producing ultraviolet radiation properly shielded?

Flammable and Combustible Materials

- Are combustible scrap, debris, and waste materials (e.g., oily rags) stored in covered metal receptacles and removed from the worksite promptly?
- Is proper storage practiced to minimize the risk of fire, including spontaneous combustion?
- Are approved containers and tanks used for the storage and handling of flammable and combustible liquids?
- Are all connections on drums and combustible liquid piping, vapor, and liquid tight?
- Are all flammable liquids kept in closed containers when not in use (e.g., parts cleaning tanks, pans)?
- Are bulk drums of flammable liquids grounded and bonded to containers during dispensing?
- Do storage rooms for flammable and combustible liquids have explosion-proof lights?
- Do storage rooms for flammable and combustible liquids have mechanical or gravity ventilation?
- Is liquefied petroleum gas stored, handled, and used in accordance with safe practices and standards?
- Are liquefied petroleum storage tanks guarded to prevent damage from vehicles?
- Are all solvent wastes and flammable liquids kept in fire-resistant covered containers until they are removed from the worksite?
- Is vacuuming used whenever possible rather than blowing or sweeping combustible dust?
- Are fire separators placed between containers of combustibles or flammables, when stacked one upon another, to ensure their support and stability?
- Are fuel-gas cylinders and oxygen cylinders separated by distance, fire-resistant barriers, or other means while in storage?
- Are fire extinguishers selected and provided for the types of materials in areas where they are to be used?
  - Class A: Ordinary combustible material fires.
  - Class B: Flammable liquid, gas, or grease fires.
  - Class C: Energized-electrical equipment fires.
- If a Halon 1301 fire extinguisher is used, can employees evacuate within the specified time for that extinguisher?
- Are appropriate fire extinguishers mounted within 75' of outside areas containing flammable liquids and within 10' of any inside storage area for such materials?
- Is the transfer/withdrawal of flammable or combustible liquids performed by trained personnel?
- Are fire extinguishers mounted so that employees do not have to travel more than 75' for a class A fire or 50' for a class B fire?
- Are employees trained in the use of fire extinguishers?
- Are extinguishers free from obstructions or blockage?
- Are all extinguishers serviced, maintained, and tagged at intervals not to exceed 1 year?

- Are all extinguishers fully charged and in their designated places?
- Is a record maintained of required monthly checks of extinguishers?
- Where sprinkler systems are permanently installed, are the nozzle heads directed or arranged so that water will not be sprayed into operating electrical switchboards and equipment?
- Are "NO SMOKING" signs posted where appropriate in areas where flammable or combustible materials are used or stored?
- Are "NO SMOKING" signs posted on liquefied petroleum gas tanks?
- Are "NO SMOKING" rules enforced in areas involving storage and use of flammable materials?
- Are safety cans used for dispensing flammable or combustible liquids at a point of use?
- Are all spills of flammable or combustible liquids cleaned up promptly?
- Are storage tanks adequately vented to prevent the development of excessive vacuum or pressure as a result of filling, emptying, or atmospheric temperature changes?
- Are storage tanks equipped with emergency venting that will relieve excessive internal pressure caused by fire exposure?
- Are spare portable or butane tanks, which are used by industrial trucks, stored in accordance with regulations?

Fire Protection

- Do you have a fire prevention plan?
- Does your plan describe the type of fire protection equipment and/or systems?
- Have you established practices and procedures to control potential fire hazards and ignition sources?
- Are employees aware of the fire hazards of the material and processes to which they are exposed?
- Is your local fire department well acquainted with your facilities, location, and specific hazards?
- If you have a fire alarm system, is it tested at least annually?
- If you have a fire alarm system, is it certified as required?
- If you have interior standpipes and valves, are they inspected regularly?
- If you have outside private fire hydrants, are they flushed at least once a year and on a routine preventive maintenance schedule?
- Are fire doors and shutters in good operating condition?
- Are fire doors and shutters unobstructed and protected against obstructions, including their counterweights?
- Are fire door and shutter fusible links in place?
- Are automatic sprinkler system water control valves and air and water pressures checked weekly/periodically as required?
- Is maintenance of automatic sprinkler system assigned to responsible persons or to a sprinkler contractor?
- Are sprinkler heads protected by metal guards when exposed to physical damage?
- Is proper clearance maintained below sprinkler heads?
- Are portable fire extinguishers provided in adequate numbers and types?
- Are fire extinguishers mounted in readily accessible locations?
- Are fire extinguishers recharged regularly and noted on the inspection tag?
- Are employees periodically instructed in the use of extinguishers and fire protection procedures?

Hazardous Chemical Exposures

- Are employees trained in the safe handling practices for hazardous chemicals such as acids, caustics, and the like?
- Are employees aware of the potential hazards involving various chemicals stored or used in the workplace—such as acids, bases, caustics, epoxys, and phenols?
- Is employee exposure to chemicals kept within acceptable levels?
- Are eyewash fountains and safety showers provided in areas where corrosive chemicals are handled?
- Are all containers, such as vats and storage tanks, labeled as to their contents, e.g., "CAUSTICS"?
- Are all employees required to use personal protective clothing and equipment when handling chemicals (i.e., gloves, eye protection, and respirators)?
- Are flammable or toxic chemicals kept in closed containers when not in use?
- Are chemical piping systems clearly marked as to their content?
- Where corrosive liquids are frequently handled in open containers or drawn from storage vessels or pipe-lines, are adequate means readily available for neutralizing or disposing of spills or overflows properly and safely?
- Have standard operating procedures been established and are they being followed when cleaning up chemical spills?
- Where needed for emergency use, are respirators stored in a convenient, clean, and sanitary location?
- Are respirators intended for emergency use adequate for the various uses for which they may be needed?
- Are employees prohibited from eating in areas where hazardous chemicals are present?
- Is personal protective equipment provided, used, and maintained whenever necessary?
- Are there written standard operating procedures for the selection and use of respirators where needed?
- If you have a respirator protection program, are your employees instructed on the correct usage and limitations of the respirators?
- Are the respirators NIOSH-approved for this particular application?
- Are they regularly inspected and cleaned, sanitized, and maintained?
- If hazardous substances are used in your processes, do you have a medical or biological monitoring system in operation?
- Are you familiar with the Threshold Limit Values or Permissible Exposure Limits of airborne contaminants and physical agents used in your workplace?
- Have control procedures been instituted for hazardous materials, where appropriate, such as respirators, ventilation systems, handling practices, and the like?
- Whenever possible, are hazardous substances handled in properly designed and exhausted booths or similar locations?
- Do you use general dilution or local exhaust ventilation systems to control dusts, vapors, gases, fumes, smoke, solvents, or mists that may be generated in your workplace?
- Is ventilation equipment provided for removal of contaminants from such operations as production grinding, buffing, spray painting, and/or vapor decreasing, and is it operating properly?
- Do employees complain about dizziness, headaches, nausea, irritation, or other factors of discomfort when they use solvents or other chemicals?
- Is there a dermatitis problem—do employees complain about skin dryness, irritation, or sensitization?

- Have you considered the use of an industrial hygienist or environmental health specialist to evaluate your operation?

- If internal combustion engines are used, is carbon monoxide kept within acceptable levels?

- Is vacuuming used, rather than blowing or sweeping whenever possible for dust cleanup?

- Are materials that give off toxic asphyxiant, suffocating, or anesthetic fumes stored in remote or isolated locations when not in use?

Hazardous Substances Communication

- Is there a list of hazardous substances used in your workplace?

- Is there a written hazard communication program dealing with Material Safety Data Sheets (MSDS) labeling and employee training?

- Who is responsible for MSDSs, container labeling, and employee training?

- Is each container for a hazardous substance (e.g., vats, bottles, storage tanks) labeled with product identity and a hazard warning (communication of the specific health hazards and physical hazards)?

- Is there a Material Safety Data Sheet readily available for each hazardous substance used?

- How will you inform other employers whose employees share the same work area where the hazardous substances are used?

- Is there an employee training program for hazardous substances?

- Does this program include:

  ○ An explanation of what an MSDS is and how to use and obtain one?

  ○ MSDS contents for each hazardous substance or class of substances?

  ○ Explanation of "Right to Know"?

  ○ Identification of where employees can see the employer's written hazard communication program and where hazardous substances are present in their work area?

  ○ The physical and health hazards of substances in the work area, how to detect their presence, and specific protective measures to be used?

  ○ Details of the hazard communication program, including how to use the labeling system and MSDSs?

  ○ How employees will be informed of hazards in nonroutine tasks and hazards of unlabeled pipes?

Electrical

- Are your workplace electricians familiar with the Cal/OSHA Electrical Safety Orders?

- Do you specify compliance with Cal/OSHA for all contract electrical work?

- Are all employees required to report as soon as practicable any obvious hazard to life or property observed in connection with electrical equipment or lines?

- Are employees instructed to make preliminary inspections and/or appropriate tests to determine what conditions exist before starting work on electrical equipment or lines?

- When electrical equipment or lines are to be serviced, maintained, or adjusted, are necessary switches opened, locked out, and tagged whenever possible?

- Are portable electrical tools and equipment grounded or of the double-insulated type?

- Are electrical appliances such as vacuum cleaners, polishers, and vending machines grounded?

- Do extension cords being used have a grounding conductor?

- Are multiple plug adapters prohibited?

- Are groundfault circuit interrupters installed on each temporary 15- or 20-ampere, 120-volt AC circuit at locations where construction, demolition, modifications, alterations, or excavations are being performed?

- Are all temporary circuits protected by suitable disconnecting switches or plug connectors at the junction with permanent wiring?

- Are exposed wiring and cords with frayed or deteriorated insulation repaired or replaced promptly?

- Are flexible cords and cables free of splices or taps?

- Are clamps or other securing means provided on flexible cords or cables at plugs, receptacles, tools, and equipment, and is the cord jacket securely held in place?

- Are all cord, cable, and raceway connections intact and secure?

- In wet or damp locations, are electrical tools and equipment appropriate for the use or location or otherwise protected?

- Is the location of electrical power lines and cables (overhead, underground, under-floor, other side of walls) determined before digging, drilling, or similar work is begun?

- Are metal measuring tapes, ropes, handlines, or similar devices with metallic thread woven into the fabric prohibited where they could come in contact with energized parts of equipment or circuit conductors?

- Is the use of metal ladders prohibited in areas where the ladder or the person using the ladder could come in contact with energized parts of equipment, fixtures, or circuit conductors?

- Are all disconnecting switches and circuit breakers labeled to indicate their use or equipment served?

- Are disconnecting means always opened before fuses are replaced?

- Do all interior wiring systems include provisions for grounding metal parts of electrical raceways, equipment, and enclosures?

- Are all electrical raceways and enclosures securely fastened in place?

- Are all energized parts of electrical circuits and equipment guarded against accidental contact by approved cabinets or enclosures?

- Is sufficient access and working space provided and maintained about all electrical equipment to permit ready and safe operations and maintenance?

- Are all unused openings (including conduit knockouts) in electrical enclosures and fittings closed with appropriate covers, plugs, or plates?

- Are electrical enclosures such as switches, receptacles, junction boxes, etc., provided with tight-fitting covers or plates?

- Are disconnecting switches for electrical motors in excess of 2 horsepower capable of opening the circuit when the motor is in a stalled condition without exploding? (Switches must be horsepower-rated equal to or in excess of the motor hp rating).

- Is low-voltage protection provided in the control device of motors driving machines or equipment that could cause probable injury from inadvertent starting?

- Is each motor disconnecting switch or circuit breaker located within sight of the motor control device?

- Is each motor located within sight of its controller or the controller disconnecting means capable of being locked in the open position, or is a separate disconnecting means installed in the circuit within sight of the motor?

- Is the controller for each motor in excess of 2 horsepower rated in horsepower equal to or in excess of the rating of the motor it serves?

- Are employees who regularly work on or around energized electrical equipment or lines instructed in cardiopulmonary resuscitation (CPR) methods?

- Are employees prohibited from working alone on energized lines or equipment over 600 volts?

Noise

- Are there areas in the workplace where continuous noise levels exceed 85 dBA? (To determine maximum allowable levels for intermittent or impact noise, see Title 8, Section 5097.)

- Are noise levels being measured using a sound-level meter or an octave band analyzer and are records being kept?

- Have you tried isolating noisy machinery from the rest of your operation?

- Have engineering controls been used to reduce excessive noise levels?

- Where engineering controls are determined not feasible, are administrative controls (i.e., worker rotation) being used to minimize individual employee exposure to noise?

- Is there an ongoing preventive health program to educate employees on safe levels of noise and exposure, effects of noise on their health, and use of personal protection?

- Is the training repeated annually for employees exposed to continuous noise above 85 dBA?

- Have work areas where noise levels make voice communication between employees difficult been identified and posted?

- Is approved hearing protective equipment (noise attenuating devices) available to every employee working in areas where continuous noise levels exceed 85 dBA?

- If you use ear protectors, are employees properly fitted and instructed in their use and care?

- Are employees exposed to continuous noise above 85 dBA given periodic audiometric testing to ensure that you have an effective hearing protection system?

Fueling

- Is it prohibited to fuel an internal combustion engine with a flammable liquid while the engine is running?

- Are fueling operations done in such a manner that likelihood of spillage will be minimal?

- When spillage occurs during fueling operations, is the spilled fuel cleaned up completely or evaporated or other measures taken to control vapors before restarting the engine?

- Are fuel tank caps replaced and secured before starting the engine?

- In fueling operations, is there always metal contact between the container and fuel tank?

- Are fueling hoses of a type designed to handle the specific type of fuel?

- Is it prohibited to handle or transfer gasoline in open containers?

- Are open lights, open flames, or sparking or arcing equipment prohibited near fueling or fuel transfer operations?

- Is smoking prohibited in the vicinity of fueling operations?

- Are fueling operations prohibited in building or other enclosed areas that are not specifically ventilated for this purpose?

- Where fueling or transfer of fuel is done through a gravity flow system, are the nozzles of the self-closing type?

Identification of Piping Systems

- When nonpotable water is piped through a facility, are outlets or taps posted to alert employees that it is unsafe and not to be used for drinking, washing, or other personal use?

- When hazardous substances are transported through aboveground piping, is each pipeline identified at points where confusion could introduce hazards to employees?
- When pipelines are identified by color painting, are all visible parts of the line so identified?
- When pipelines are identified by color-painted bands or tapes, are the bands or tapes located at reasonable intervals and at each outlet, valve, or connection?
- When pipelines are identified by color, is the color code posted at all locations where confusion could introduce hazards to employees?
- When the contents of pipelines are identified by name or name abbreviation, is the information readily visible on the pipe near each valve or outlet?
- When pipelines carrying hazardous substances are identified by tags, are the tags constructed of durable materials, the message carried clearly and permanently distinguishable, and are tags installed at each valve or outlet?
- When pipelines are heated by electricity, steam, or other external source, are suitable warning signs or tags placed at unions, valves, or other serviceable parts of the system?

Materials Handling

- Is there safe clearance for equipment through aisles and doorways?
- Are aisleways designated, permanently marked, and kept clear to allow unhindered passage?
- Are motorized vehicles and mechanized equipment inspected daily or prior to use?
- Are vehicles shut off and brakes set prior to loading or unloading?
- Are containers or combustibles or flammables, when stacked while being moved, always separated by dunnage sufficient to provide stability?
- Are dock boards (bridge plates) used when loading or unloading operations are taking place between vehicles and docks?
- Are trucks and trailers secured from movement during loading and unloading operations?
- Are dock plates and loading ramps constructed and maintained with sufficient strength to support imposed loading?
- Are hand trucks maintained in safe operating condition?
- Are chutes equipped with sideboards of sufficient height to prevent the materials being handled from falling off?
- Are chutes and gravity roller sections firmly placed or secured to prevent displacement?
- At the delivery end of rollers or chutes, are provisions made to brake the movement of the handled materials?
- Are pallets usually inspected before being loaded or moved?
- Are hooks with safety latches or other arrangements used when hoisting materials so that slings or load attachments won't accidentally slip off the hoist hooks?
- Are securing chains, ropes, chockers, or slings adequate for the job to be performed?
- When hoisting material or equipment, are provisions made to ensure no one will be passing under the suspended loads?
- Are Material Safety Data Sheets available to employees handling hazardous substances?

Transporting Employees and Materials

- Do employees who operate vehicles on public thoroughfares have valid operator's licenses?

- When seven or more employees are regularly transported in a van, bus, or truck, is the operator's license appropriate for the class of vehicle being driven?

- Is each van, bus, or truck used regularly to transport employees equipped with an adequate number of seats?

- When employees are transported by truck, are there provisions to prevent their falling from the vehicle?

- Are vehicles used to transport employees equipped with lamps, brakes, horns, mirrors, windshields, and turn signals in good repair?

- Are transport vehicles provided with handrails, steps, stirrups, or similar devices so placed and arranged that employees can safely mount or dismount?

- Are employee transport vehicles equipped at all times with at least two reflective-type flares?

- Is a fully charged fire extinguisher, in good condition, with at least a 4 B:C rating maintained in each employee transport vehicle?

- When cutting tools with sharp edges are carried in passenger compartments of employee transport vehicles, are they placed in closed boxes or containers that are secured in place?

- Are employees prohibited from riding on top of any load that can shift, topple, or otherwise become unstable?

Control of Harmful Substances by Ventilation

- Is the volume and velocity of air in each exhaust system sufficient to gather the dusts, fumes, mists, vapors, or gases to be controlled and to convey them to a suitable point of disposal?

- Are exhaust inlets, ducts, and plenums designed, constructed, and supported to prevent collapse or failure of any part of the system?

- Are clean-out ports or doors provided at intervals not to exceed 12' in all horizontal runs of exhaust ducts?

- Where two or more different types of operations are being controlled through the same exhaust system, will the combination of substances being controlled constitute a fire, explosion, or chemical reaction hazard in the duct?

- Is adequate makeup air provided to areas where exhaust systems are operating?

- Is the intake for makeup air located so that only clean, fresh air that is free of contaminates will enter the work environment?

- Where two or more ventilation systems are serving a work area, is their operation such that one will not offset the functions of the other?

Sanitizing Equipment and Clothing

- Is personal protective clothing or equipment that employees are required to wear or use of a type capable of being easily cleaned and disinfected?

- Are employees prohibited from interchanging personal protective clothing or equipment unless it has been properly cleaned?

- Are machines and equipment that process, handle, or apply materials which could be injurious to employees cleaned and/or decontaminated before being overhauled or placed in storage?

- Are employees prohibited from smoking or eating in any area where contaminates are present that could be injurious if ingested?

- When employees are required to change from street clothing into protective clothing, is a clean change-room with separate storage facility for street and protective clothing provided?

- Are employees required to shower and wash their hair as soon as possible after a known contact has occurred with a carcinogen?

- When equipment, materials, or other items are taken into or removed from a carcinogen-regulated area, is it done in a manner that will not contaminate nonregulated areas or the external environment?

Tire Inflation

- Where tires are mounted and/or inflated on drop center wheels, is a safe-practice procedure posted and enforced?
- Where tires are mounted and/or inflated on wheels with split rims and/or retainer rings, is a safe-practice procedure posted and enforced?
- Does each tire inflation hose have a clip-on chuck with at least 24" of hose between the chuck and an in-line hand valve and gauge?
- Does the tire inflation control valve automatically shut off the airflow when the valve is released?
- Is a tire restraining device such as a cage, rack, or other effective means used while inflating tires mounted on split rims or rims using retainer rings?
- Are employees strictly forbidden from taking a position directly over or in front of a tire while it's being inflated?

Emergency Action Plan

- Are you required to have an emergency action plan?
- Does the emergency action plan comply with requirements of T8CCR 3220(a)?
- Have emergency escape procedures and routes been developed and communicated to all employers?
- Do employees who remain to operate critical plant operations before they evacuate know the proper procedures?
- Is the employee alarm system that provides a warning for emergency action recognizable and perceptible above ambient conditions?
- Are alarm systems properly maintained and tested regularly?
- Is the emergency action plan reviewed and revised periodically?
- Do employees know their responsibilities:
  - For reporting emergencies?
  - During an emergency?
  - For conducting rescue and medical duties?

Infection Control

- Are employees potentially exposed to infectious agents in body fluids?
- Have occasions of potential occupational exposure been identified and documented?
- Has a training and information program been provided for employees exposed to or potentially exposed to blood and/or body fluids?
- Have infection control procedures been instituted where appropriate, such as ventilation, universal precautions, workplace practices, and personal protective equipment?
- Are employees aware of specific workplace practices to follow when appropriate (e.g., hand washing, handling sharp instruments, handling of laundry, disposal of contaminated materials, reusable equipment)?
- Is personal protective equipment provided to employees and in all appropriate locations?
- Is the necessary equipment (e.g., mouthpieces, resuscitation bags, and other ventilation devices) provided for administering mouth-to-mouth resuscitation on potentially infected patients?

- Are facilities/equipment to comply with workplace practices available, such as hand-washing sinks, biohazard tags and labels, needle containers, detergents/disinfectants to clean up spills?
- Are all equipment and environmental and working surfaces cleaned and disinfected after contact with blood or potentially infectious materials?
- Is infectious waste placed in closable, leakproof containers, bags, or puncture-resistant holders with proper labels?
- Has medical surveillance, including HBV evaluation, antibody testing, and vaccination, been made available to potentially exposed employees?
- Training on universal precautions?
- Training on personal protective equipment?
- Training on workplace practices, which should include blood drawing, room cleaning, laundry handling, clean up of blood spills?
- Training on needlestick exposure/management?
- Hepatitis B vaccinations?

Ergonomics

- Can the work be performed without eyestrain or glare to the employees?
- Does the task require prolonged raising of the arms?
- Do the neck and shoulders have to be stooped to view the task?
- Are there pressure points on any parts of the body (wrists, forearms, back of thighs)?
- Can the work be done using the larger muscles of the body?
- Can the work be done without twisting or overly bending the lower back?
- Are there sufficient rest breaks, in addition to the regular rest breaks, to relieve stress from repetitive-motion tasks?
- Are tools, instruments, and machinery shaped, positioned, and handled so that tasks can be performed comfortably?
- Are all pieces of furniture adjusted, positioned, and arranged to minimize strain on all parts of the body?

Ventilation for Indoor Air Quality

- Does your HVAC system provide at least the quantity of outdoor air required by the State Building Standards Code, Title 24, Part 2 at the time the building was constructed?
- Is the HVAC system inspected at least annually and problems corrected?
- Are inspection records retained for at least 5 years?

Crane Checklist

- Are the cranes visually inspected for defective components prior to the beginning of any work shift?
- Are all electrically operated cranes effectively grounded?
- Has a crane preventive maintenance program been established?
- Is the load chart clearly visible to the operator?
- Are operating controls clearly identified?
- Is a fire extinguisher provided at the operator's station?
- Is the rated capacity visibly marked on each crane?
- Is an audible warning device mounted on each crane?

- Is sufficient illumination provided for the operator to perform the work safely?
- Are cranes of such design that the boom could fall over backward equipped with boomstops?
- Does each crane have a certificate indicating that required testing and examinations have been performed?
- Are crane inspection and maintenance records maintained and available for inspection?

Hazard Assessment and Correction Record

- Date of Inspection: Person Conducting Inspection:

- Unsafe Condition or Work Practice:

- Corrective Action Taken:

- Date of Inspection: Person Conducting Inspection:

- Unsafe Condition or Work Practice:

- Corrective Action Taken:

- Date of Inspection:

- Person Conducting Inspection:

- Unsafe Condition or Work Practice:

- Corrective Action Taken:

Accident/Exposure Investigation Report

- Date and Time of Accident:

- Location:

- Accident Description:

- Employees Involved:

- Preventive Action Recommendations:

- Corrective Actions Taken:

- Manager Responsible: Date Completed:

Worker Training and Instruction Record

Employee Name	Training Dates	Type of Training	Trainers

JHA—COMMON HAZARDS AND DESCRIPTIONS

The following are excerpts from OSHA Publication 3071. The full version of this publication can be found at: http://www.osha.gov. This information can also be found at dewalt.com/guides.

Hazards	Hazard Descriptions
Chemical (Toxic)	A chemical that exposes a person by absorption through the skin, inhalation, or through the bloodstream that causes illness, disease, or death. The amount of chemical exposure is critical in determining hazardous effects. Check Material Safety Data Sheets (MSDS) and/or OSHA 1910.1000 for chemical hazard information.
Chemical (Flammable)	A chemical that, when exposed to a heat ignition source, results in combustion. Typically, the lower a chemical's flash point and boiling point, the more flammable the chemical. Check MSDS for flammability information.
Chemical (Corrosive)	A chemical that, when it comes into contact with skin, metal, or other materials, damages the materials. Acids and bases are examples of corrosives.
Explosion (Chemical Reaction)	Self-explanatory.
Explosion (Over Pressurization)	Sudden and violent release of a large amount of gas/energy due to a significant pressure difference such as rupture in a boiler or compressed gas cylinder.
Electrical (Shock/Short Circuit)	Contact with exposed conductors or a device that is incorrectly or inadvertently grounded, such as when a metal ladder comes into contact with power lines. 60 Hz alternating current (common house current) is very dangerous because it can stop the heart.
Electrical (Fire)	Use of electrical power that results in electrical overheating or arcing to the point of combustion or ignition of flammables or electrical component damage.

(continues)

Electrical (Static/ESD)	The moving or rubbing of wool, nylon, other synthetic fibers, and even flowing liquids can generate static electricity. This creates an excess or deficiency of electrons on the surface of material that discharges (spark) to the ground, resulting in the ignition of flammables or damage to electronics or the body's nervous system.
Electrical (Loss of Power)	Safety-critical equipment failure as a result of loss of power.
Ergonomics (Strain)	Damage of tissue due to overexertion (strains and sprains) or repetitive motion.
Ergonomics (Human Error)	A system design, procedure, or equipment that is error-provocative. (A switch goes up to turn something off.)
Excavation (Collapse)	Soil collapse in a trench or excavation as a result of improper or inadequate shoring. Soil type is critical in determining the hazard likelihood.
Fall (Slip, Trip)	Conditions that result in falls (impacts) from height or traditional walking surfaces (such as slippery floors, poor housekeeping, uneven walking surfaces, exposed ledges, etc.)
Fire/Heat	Temperatures that can cause burns to the skin or damage to other organs. Fires require a heat source, fuel, and oxygen.
Mechanical/Vibration (Chaffing/Fatigue)	Vibration that can cause damage to nerve endings, or material fatigue that results in a safety-critical failure. (Examples are abraded slings and ropes, weakened hoses and belts.)
Mechanical Failure	Self-explanatory; typically occurs when devices exceed designed capacity or are inadequately maintained.
Mechanical	Skin, muscle, or body part exposed to crushing or caught between cutting, tearing, shearing items, or equipment.
Noise	Noise levels (>85 dBA 8 hr TWA) that result in hearing damage or inability to communicate safety-critical information.
Radiation (Ionizing)	Alpha, beta, gamma, neutral particles, and X-rays that cause injury (tissue damage) by ionization of cellular components.
Radiation (Nonionizing)	Ultraviolet, visible light, infrared, and microwaves that cause injury to tissue by thermal or photochemical means.
Struck By (Mass Acceleration)	Accelerated mass that strikes the body causing injury or death. (Examples are falling objects and projectiles.)
Struck Against	Injury to a body part as a result of coming into contact of a surface in which action was initiated by the person. (An example is when a screwdriver slips.)
Temperature Extreme (Heat/Cold)	Temperatures that result in heat stress, exhaustion, or metabolic slowdown such as hypothermia.
Visibility	Lack of lighting or obstructed vision that results in an error or other hazard.
Weather Phenomena (Snow/Rain/Wind/Ice)	Self-explanatory.

SAMPLE JOB HAZARD ANALYSIS (JHA) FORMS

The following forms help you to provide documentation around job hazard analysis and training/communication. These forms can be found at dewalt.com/guides. They are based on OSHA's Job Hazard Analysis Publication 3071, which can be found at http://www.osha.gov.

ANALYZING JOB HAZARDS

Contractor: _____   Project Name & Number: _____   Date: _____

Job Location: _____   Superintendent/Foreman: _____   Analysis by:_____

Work Description:_____   Required PPE: _____

Work Activity	Potential Hazards	Safe Work Practices

JHA TRAINING/COMMUNICATION

Training Date: _____   Training Conducted By: _____

Employee Printed Name	Employee Number	Employee Signature

JHA—HAZARD CONTROL MEASURES

The following are excerpts from OSHA Publication 3071. This information can also be found at dewalt.com/guides. The full version of this publication can be found at: http://www.osha.gov.

Hazard Control Measures

Information obtained from a job hazard analysis is useless unless hazard control measures recommended in the analysis are incorporated into the tasks. Managers should recognize that not all hazard controls are equal. Some are more effective than others at reducing the risk.

The order of precedence and effectiveness of hazard control is the following:

1. Engineering controls.
2. Administrative controls.
3. Personal protective equipment.

Engineering controls include the following:

- Elimination/minimization of the hazard—Designing the facility, equipment, or process to remove the hazard, or substituting processes, equipment, materials, or other factors to lessen the hazard;
- Enclosure of the hazard using enclosed cabs, enclosures for noisy equipment, or other means;
- Isolation of the hazard with interlocks, machine guards, blast shields, welding curtains, or other means; and
- Removal or redirection of the hazard such as with local and exhaust ventilation.

Administrative controls include the following:

- Written operating procedures, work permits, and safe work practices;
- Exposure time limitations (used most commonly to control temperature extremes and ergonomic hazards);
- Monitoring the use of highly hazardous materials;
- Alarms, signs, and warnings;
- Buddy system; and
- Training.

Personal Protective Equipment—such as respirators, hearing protection, protective clothing, safety glasses, and hardhats—is acceptable as a control method in the following circumstances:

- When engineering controls are not feasible or do not totally eliminate the hazard;
- While engineering controls are being developed;
- When safe work practices do not provide sufficient additional protection; and
- During emergencies when engineering controls may not be feasible.

Use of one hazard control method over another higher in the control precedence may be appropriate for providing interim protection until the hazard is abated permanently. In reality, if the hazard cannot be eliminated entirely, the adopted control measures will likely be a combination of all three items instituted simultaneously.

OSHA SAMPLE SAFETY MEETING TOPICS

These excerpts are just a few of the OSHA quick cards available for reference at http://www.osha.gov. These samples can also be found at dewalt.com/guides.

Aerial Lifts
Protect Yourself

Aerial lifts include boom-supported aerial platforms, such as cherry pickers or bucket trucks. The major causes of fatalities are falls, electrocutions and collapses or tip overs.

Safe Work Practices

- Make sure that workers who operate aerial lifts are properly trained in the safe use of the equipment.
- Maintain and operate elevating work platforms according to the manufacturer's instructions.
- Never override hydraulic, mechanical, or electrical safety devices.
- Never move the equipment with workers in an elevated platform unless this is permitted by the manufacturer.
- Do not allow workers to position themselves between overhead hazards, such as joists and beams, and the rails of the basket. Movement of the lift could crush the worker(s).
- Maintain a minimum clearance of at least 10 feet, or 3 meters, away from the nearest overhead lines.
- Always treat power lines, wires and other conductors as energized, even if they are down or appear to be insulated.
- Use a body harness or restraining belt with a lanyard attached to the boom or basket to prevent the worker(s) from being ejected or pulled from the basket.
- Set the brakes and use wheel chocks when on an incline.
- Use outriggers, if provided.
- Do not exceed the load limits of the equipment. Allow for the combined weight of the worker, tools and materials.

For more complete information:

Occupational Safety and Health Administration
U.S. Department of Labor
www.osha.gov (800) 321-OSHA

OSHA 3267-09N-05

Jirafas
Protéjase

Las jirafas incluyen a las plataformas elevadas de aguilón sostenido, como lo son las de puntal extensible con canasta (mejor conocidas como "cherry pickers") o los camiones canasta. Las principales causas de muertes son por caídas, electrocuciones y colapsos o volteos.

Prácticas de Trabajo Seguras

- Asegúrese que los trabajadores que operan jirafas están adecuadamente adiestrados en el uso seguro del equipo.
- Mantenga y opere las plataformas de trabajo elevadas de acuerdo con las instrucciones del manufacturero.
- Nunca invalide los dispositivos de seguridad hidráulicos, mecánicos o eléctricos.
- Nunca mueva el equipo con trabajadores en una plataforma elevada, a menos que sea permitido por el manufacturero.
- No permita a los trabajadores ponerse entre riesgos que estén por encima de la cabeza, como viguetas y vigas, y las barandas del canasto. El movimiento de la jirafa puede aplastar al trabajador.
- Mantenga una distancia mínima segura de las líneas eléctricas aéreas más cercanas de al menos 10 pies, ó 3 metros.
- Siempre trate a las líneas de energía eléctrica, alambres y otros conductores como si estuvieran energizados (vivos), aún si están fuera de servicio o parece que están aislados.
- Use un arnés de cuerpo o correa que restringe el movimiento con una cuerda de seguridad atada al aguilón o canasto para prevenir que el trabajador salga disparado o sea tirado del canasto.
- Ponga los frenos y use calzos cuando esté en un área inclinada.
- Use estabilizadores, si son provistos.
- No exceda la carga límite del equipo. Tome en cuenta el peso combinado del trabajador, herramientas y materiales.

Para información más completa:

Administración de Seguridad y Salud Ocupacional
Departamento del Trabajo de EE.UU.
www.osha.gov (800) 321-OSHA

Courtesy of the Occupational Safety & Health Administration, U.S. Department of Labor

OSHA® QUICK CARD™

Pneumatic Nail Gun
Safety

Nail guns drive nails and staples into building materials. Injuries or fatalities can result from improper use.

Hazards

The operator and coworkers are at risk. Eyes, hands and fingers are especially at risk. Nails can:

• Splinter or blow out fragments from the material.
• Puncture the back of the material.
• Fire completely through the material and strike workers behind the nailing surface.
• Pose contact hazards, such as nails striking electrical wires.

Common Tool Types

Sequential tools: Require nose (workpiece contact) to be depressed before the trigger is pulled.

• Two-step sequence makes accidental firing less likely.
• Can be used for most nailing tasks.

Contact tools: Tool fires anytime the trigger and nose (workpiece contact) are both depressed.

• Trigger can be held down to allow "bump firing".
• Use for nailing on flat surfaces.
  CAUTION:
  ▪ If trigger is depressed, the tool will fire anytime the nose is depressed; can also cause unwanted double firing of nails.

Safe Work Practices and PPE

• Follow manufacturer's tool labels and operating manual.
• Wear safety glasses with side shields.
• Never defeat or modify safety features.
• Keep fingers away from trigger when not driving nails.
• Sequential tools have reduced risk of accidental and double firing.
• Avoid line of fire hazards in front of and behind material; position yourself (especially your free hand) out of the line of fire. Never point nail gun at anyone. Watch for coworkers behind the nailing surface.
• Disconnect the gun to perform maintenance, move to another work area, or clear jams.
• Train on safe operating procedures, proper body placement and correct PPE use.
• NOTE: The tool must meet applicable OSHA guarding standards.

OSHA® Occupational
Safety and Health
Administration
U.S. Department of Labor
www.osha.gov   (800) 321-OSHA

OSHA 3363-4-09

OSHA® DATOS RÁPIDOS

Pistolas Neumáticas de Clavos
Seguridad

Las pistolas de clavos insertan clavos y grapas en materiales de construcción. Su uso inapropiado puede resultar en lesiones o muertes.

Riesgos

El operador y sus compañeros de trabajo están en riesgo. Particularmente en los ojos, manos y dedos. Los clavos/grapas pueden:

• Astillar o expulsar fragmentos del material.
• Perforar el material por su lado posterior.
• Disparar y atravesar completamente el material y golpear a los trabajadores al otro lado de la superficie sobre la que se está clavando.
• Presentar riesgos de contacto, como clavos golpeando cables eléctricos.

Tipos comunes de herramientas

Herramientas secuenciales: Requieren que su punta (la parte que hace contacto con la pieza de trabajo) sea presionada antes de que se hale el gatillo.

• Secuencia de dos pasos reduce la probabilidad de disparos accidentales.
• Pueden utilizarse para la mayoría de las tareas que involucran clavar.

Herramientas de contacto: La herramienta dispara siempre que se presione el gatillo y la punta (la parte que hace contacto con la pieza de trabajo)

• El gatillo puede presionarse hacia abajo para permitir "disparos repetidos"
• Se utilizan para clavar sobre superficies planas.
• Si se ha presionado el gatillo, la herramienta disparará en cualquier momento que la punta esté presionada; también puede provocar el disparo doble involuntario de los clavos.

Prácticas seguras de trabajo y PPE

• Siga las instrucciones en las etiquetas del manufacturero en la herramienta y su manual de operación.
• Utilice gafas de protección con escudos laterales.
• Nunca anule o modifique los aditamentos de seguridad.
• Mantenga los dedos alejados del gatillo cuando no esté insertando clavos.
• Las herramientas secuenciales han reducido el riesgo de disparos dobles y accidentales.
• Evite los riesgos inherentes de colocarse en la línea de disparo en frente o detrás del material sobre el que esté clavando; debe posicionarse fuera de la línea de disparo (especialmente su mano libre). Nunca apunte a nadie con la pistola. Esté pendiente de la presencia de trabajadores detrás de la superficie sobre la cual se esté clavando.
• Desconecte la pistola al brindarle mantenimiento, trasladarse a otra área de trabajo o despejar atascamientos.
• Adiéstrese sobre los procedimientos seguros de operación, las apropiadas posturas corporales y el uso correcto de PPE.
• NOTA: La herramienta debe cumplir con las normas aplicables de resguardo de OSHA.

Para información más completa:

OSHA® Administración de
Seguridad y Salud
Ocupacional
Departamento del Trabajo de EE.UU.
www.osha.gov (800) 321-OSHA

OSHA® QUICK CARD™

Chipper Machine Safety

Chipper machines cut tree limbs into small chips. Hazards arise when operators get too close to, or make contact with, the chipper. Contact with chipper operating components (blades, discs or knives) may result in amputation or death. Operators may also be injured by material thrown from the machine. To minimize these hazards, use appropriate engineering and work practice controls, including operator training.

Hazards

* Operators making contact with or being pulled into the chipper.
* Hearing loss.
* Face, eye, head or hand injuries.

Safe Work Practices

* **Never reach into a chipper while it is operating.**
* Do not wear loose-fitting clothing around a chipper.
* Always follow the manufacturer's guidelines and safety instructions.
* Use earplugs, safety glasses, hard hats and gloves.
* Workers should be trained on the safe operation of chipper machines. Always supervise new workers using a chipper to ensure that they work safely and never endanger themselves or others.
* Protect yourself from contacting operating chipper components by guarding the infeed and discharge ports, and preventing the opening of the access covers or doors until the drum or disc completely stops.
* Prevent detached trailer chippers from rolling or sliding on slopes by chocking the trailer wheels.
* Maintain a safe distance (i.e., two tree or log lengths) between chipper operations and other tree work or workers.
* When servicing and/or maintaining chipping equipment (i.e., "unjamming") use a lockout system to ensure that the equipment is de-energized.

For more complete information:

OSHA® Occupational Safety and Health Administration
U.S. Department of Labor
www.osha.gov   (800) 321-OSHA

OSHA 3279-05R-07

OSHA® DATOS RÁPIDOS

Astilladoras Seguridad

Las astilladoras cortan ramas de árboles en pedazos pequeños. Los riesgos surgen cuando los operarios se acercan demasiado o hacen contacto con los astilladora. El contacto con los componentes de una astilladora en funcionamiento (cuchillas, discos o cuchillos) puede tener como resultado una amputación o la muerte. Los operarios también pueden ser lesionados por el material que sale de la máquina. Para reducir estos riesgos, use los controles de ingeniería y prácticas de trabajo apropiados, incluyendo la capacitación de los operarios.

Riesgos

* Operarios teniendo contacto o siendo halados hacia dentro de la astilladora.
* Pérdida de audición.
* Lesiones en la cara, ojos, cabeza o manos.

Prácticas de Trabajo Seguras

* **Nunca ponga las manos adentro de una astilladora mientras esté en funcionamiento.**
* No use ropa holgada alrededor de una astilladora.
* Siga siempre las recomendaciones e instrucciones de seguridad del manufacturero.
* Use tapones para oídos, gafas de seguridad, casco de seguridad y guantes.
* Los trabajadores deben estar adiestrados en la operación segura de las astilladoras. Siempre supervise a los trabajadores nuevos que usen astilladoras para asegurarse que ellos trabajan de manera segura y que nunca se ponen en peligro o ponen en peligro a otros.
* Protéjase de contactar los componentes de astilladoras en funcionamiento poniendo guardas en los puntos de alimentación y descarga, y evitando que se abran las cubiertas o puertas de acceso hasta que el tambor o disco se detenga completamente.
* Evite que remolques desconectados de las astilladoras rueden o se deslicen en áreas inclinadas poniéndole calzos a las ruedas del remolque.
* Mantenga una distancia segura (o sea, el largo de 2 árboles o troncos) entre trabajos con astilladoras y otros trabajos en árboles/trabajadores.
* Cuando se esté dando servicio o mantenimiento a equipo de astillar (o sea, desatascándolo) use el sistema de candado para asegurarse de que el equipo esté desactivado.

Para información más completa:

OSHA® Administración de Seguridad y Salud Ocupacional
Departamento del Trabajo de EE.UU.
www.osha.gov (800) 321-OSHA

SAFETY MEETING ROSTER

This sample safety meeting roster can also be found at dewalt.com/guides.

Company Name: _____   Job Name: _____

Date: _____   Presented By: _____

Topic(s): _____

Print Name	Signature	Company Name

Courtesy Dan Johnson

Comments: _____

_____

_____

_____

SAMPLE FALL PROTECTION PLAN FOR PRECAST CONCRETE ERECTION

OSHA's sample fall protection plan for precast concrete erection can be found at http://www.osha.gov, as well as dewalt.com/guides.

FALL PROTECTION PLAN FOR PRECAST/PRE-STRESSED CONCRETE STRUCTURES

This Fall Protection Plan is specific for the following project:

Location of Job:

Erecting Company:

Date Plan Prepared or Modified:

Plan Prepared By:

Plan Approved By:

Plan Supervised By:

The following Fall Protection Plan is a sample program prepared for the prevention of injuries associated with falls. A Fall Protection Plan must be developed and evaluated on a site-by-site basis. It is recommended that erectors discuss the written Fall Protection Plan with their OSHA Area Office prior to going on a jobsite.

I. Statement of Company Policy

(Company Name) is dedicated to the protection of its employees from on-the-job injuries. All employees of (Company Name) have the responsibility to work safely on the job. The purpose of this plan is: (a) To supplement our standard safety policy by providing safety standards specifically designed to cover fall protection on this job and (b) to ensure that each employee is trained and made aware of the safety provisions which are to be implemented by this plan prior to the start of erection.

This Fall Protection Plan addresses the use of other than conventional fall protection at a number of areas on the project, as well as identifying specific activities that require nonconventional means of fall protection. These areas include:

a. Connecting activity (point of erection).
b. Leading edge work.
c. Unprotected sides or edge.
d. Grouting.

This plan is designed to enable employers and employees to recognize the fall hazards on this job and to establish the procedures that are to be followed in order to prevent falls to lower levels or through holes and openings in walking/working surfaces. Each employee will be trained in these procedures and strictly adhere to them except when doing so would expose the employee to a greater hazard. If, in the employee's opinion, this is the case, the employee is to notify the foreman of the concern and the concern addressed before proceeding.

Safety policy and procedure on any one project cannot be administered, implemented, monitored, and enforced by any one individual. The total objective of a safe, accident-free work environment can only be accomplished by a dedicated, concerted effort by every individual involved with the project from management down to the last employee. Each employee must understand their value to the company; the costs of accidents, both monetary, physical, and emotional; the objective of the safety policy and procedures; the safety rules that apply to the safety policy and procedures; and what their individual role is in administering, implementing, monitoring, and compliance to their safety policy and procedures. This allows for a more personal approach to compliance through planning, training, understanding, and cooperative effort, rather than by strict enforcement. If for any reason an unsafe act persists, strict enforcement will be implemented.

It is the responsibility of (name of competent person) to implement this Fall Protection Plan. (Name of Competent Person) is responsible for continual observational safety checks of their work operations and to enforce the safety policy and procedures. The foreman also is responsible for correcting any unsafe acts or conditions immediately. It is the responsibility of the employee to understand and adhere to the procedures of this plan and to follow the instructions of the foreman. It is also the responsibility of the employee to bring to management's attention any unsafe or hazardous conditions or acts that may cause injury to either themselves or any other employees. Any changes to this Fall Protection Plan must be approved by (name of Qualified Person).

II. Fall Protection Systems to be Used on this Project

Where conventional fall protection is infeasible or creates a greater hazard at the leading edge and during initial connecting activity, we plan to do this work using a safety monitoring system and expose only a minimum number of employees for the time necessary to actually accomplish the job. The maximum number of workers to be monitored by one safety monitor is six (6). We are designating the following trained employees as designated erectors and they are permitted to enter the controlled access zones and work without the use of conventional fall protection.

Safety monitor:

Designated erector:

Designated erector:

Designated erector:

Designated erector:

Designated erector:

Designated erector:

The safety monitor shall be identified by wearing an orange hard hat. The designated erectors will be identified by one of the following methods:

1. They will wear a blue-colored armband, or

2. They will wear a blue-colored hard hat, or

3. They will wear a blue-colored vest.

Only individuals with the appropriate experience, skills, and training will be authorized as designated erectors. All employees that will be working as designated erectors under the safety monitoring system shall have been trained and instructed in the following areas:

1. Recognition of the fall hazards in the work area (at the leading edge and when making initial connections-point of erection).

2. Avoidance of fall hazards using established work practices which have been made known to the employees.

3. Recognition of unsafe practices or working conditions that could lead to a fall, such as windy conditions.

4. The function, use, and operation of safety monitoring systems, guardrail systems, body belt/harness systems, control zones, and other protection to be used.

5. The correct procedure for erecting, maintaining, disassembling, and inspecting the system(s) to be used.

6. Knowledge of construction sequence or the erection plan.

A conference will take place prior to starting work involving all members of the erection crew, crane crew, and supervisors of any other concerned contractors. This conference will be conducted by the precast concrete erection supervisor in charge of the project. During the prework conference, erection procedures and sequences pertinent to this job will be thoroughly discussed and safety practices to be used throughout the project will be specified. Further, all personnel will be informed that the controlled access zones are off-limits to all personnel other than those designated erectors specifically trained to work in that area.

Safety Monitoring System

A safety monitoring system means a fall protection system in which a competent person is responsible for recognizing and warning employees of fall hazards. The duties of the safety monitor are to:

1. Warn by voice when approaching the open edge in an unsafe manner.

2. Warn by voice if there is a dangerous situation developing which cannot be seen by another person involved with product placement, such as a member getting out of control.

3. Make the designated erectors aware they are in a dangerous area.

4. Be competent in recognizing fall hazards.

5. Warn employees when they appear to be unaware of a fall hazard or are acting in an unsafe manner.

6. Be on the same walking/working surface as the monitored employees and within visual sighting distance of the monitored employees.

7. Be close enough to communicate orally with the employees.

8. Not allow other responsibilities to encumber monitoring. If the safety monitor becomes too encumbered with other responsibilities, the monitor shall (1) stop the erection process; and (2) turn over other responsibilities to a designated erector; or (3) turn over the safety monitoring function to another designated, competent person.

The safety monitoring system shall not be used when the wind is strong enough to cause loads with large surface areas to swing out of radius, or result in loss of control of the load, or when weather conditions cause the walking/working surfaces to become icy or slippery.

Control Zone System

A controlled access zone means an area designated and clearly marked in which leading edge work may take place without the use of guardrail, safety net, or personal fall arrest systems to protect the employees in the area. Control zone systems shall comply with the following provisions:

1. When used to control access to areas where leading edge and other operations are taking place, the controlled access zone shall be defined by a control line or by any other means that restricts access. When control lines are used, they shall be erected not less than 6' nor more than 60' or half the length of the member being erected, whichever is less, from the leading edge.
2. The control line shall extend along the entire length of the unprotected or leading edge and shall be approximately parallel to the unprotected or leading edge.
3. The control line shall be connected on each side to a guardrail system or wall.
4. Control lines shall consist of ropes, wires, tapes, or equivalent materials, and supporting stanchions as follows:
   ◦ Each line shall be flagged or otherwise clearly marked at not more than 6' intervals with high-visibility material.
   ◦ Each line shall be rigged and supported in such a way that its lowest point (including sag) is not less than 39" from the walking/working surface and its highest point is not more than 45" from the walking/working surface.
   ◦ Each line shall have a minimum breaking strength of 200 pounds.

Holes

All openings greater than 12"×12" will have perimeter guarding or covering. All predetermined holes will have the plywood covers made in the precaster's yard and shipped with the member to the jobsite. Prior to cutting holes on the job, proper protection for the hole must be provided to protect the workers. Perimeter guarding or covers will not be removed without the approval of the erection foreman.

Precast concrete column erection through the existing deck requires that many holes be provided through this deck. These are to be covered and protected. Except for the opening being currently used to erect a column, all opening protection is to be left undisturbed. The opening being uncovered to erect a column will become part of the point of erection and will be addressed as part of this Fall Protection Plan. This uncovering is to be done at the erection foreman's direction and will only occur immediately prior to "feeding" the column through the opening. Once the end of the column is through the slab opening, there will no longer exist a fall hazard at this location.

III. Implementation of Fall Protection Plan

The structure being erected is a multistory total precast concrete building consisting of columns, beams, wall panels, and hollow-core slabs and double tee floor and roof members.

The following is a list of the products and erection situations on this job:

• Columns
   ◦ For columns 10' to 36' long, employees disconnecting crane hooks from columns will work from a ladder and wear a body belt/harness with lanyard and be tied off when both hands are needed to disconnect. For tying off, a vertical lifeline will be connected to the lifting eye at the top of the column, prior to lifting, to be used with a manually operated or mobile rope grab. For columns

too high for the use of a ladder, 36' and higher, an added cable will be used to reduce the height of the disconnecting point so that a ladder can be used. This cable will be left in place until a point in erection that it can be removed safely. In some cases, columns will be unhooked from the crane by using an erection tube or shackle with a pull pin that is released from the ground after the column is stabilized.

- ○ The column will be adequately connected and/or braced to safely support the weight of a ladder with an employee on it.

- Inverted Tee Beams

  - ○ Employees erecting inverted tee beams, at a height of 6' to 40', will erect the beam and make initial connections and final alignment from a ladder. If the employee needs to reach over the side of the beam to bar or make an adjustment to the alignment of the beam, he will mount the beam and be tied off to the lifting device in the beam after ensuring the load has been stabilized on its bearing. To disconnect the crane from the beam, an employee will stand a ladder against the beam. Because the use of ladders is not practical at heights above 40', beams will be initially placed with the use of tag lines and their final alignment made by a person on a manlift or similar employee-positioning systems.

- Spandrel Beams

  - ○ Spandrel beams at the exterior of the building will be aligned as closely as possible with the use of tag lines with the final placement of the spandrel beam made from a ladder at the open end of the structure. A ladder will be used to make the initial connections and to disconnect the crane. The other end of the beam will be placed by the designated erector from the double tee deck under the observation of the safety monitor.

  - ○ The beams will be adequately connected and/or braced to safely support the weight of a ladder with an employee on it.

- Floor and Roof Members

  - ○ During installation of the precast concrete floor and/or roof members, the work deck continuously increases in area as more and more units are being erected and positioned. Thus, the unprotected floor/roof perimeter is constantly modified with the leading edge changing location as each member is installed. The fall protection for workers at the leading edge shall be assured by properly constructed and maintained control zone lines not more than 60' away from the leading edge, supplemented by a safety monitoring system to ensure the safety of all designated erectors working within the area defined by the control zone lines.

  - ○ The hollow-core slabs erected on the masonry portion of the building will be erected and grouted using the safety monitoring system. Grout will be placed in the space between the end of the slab and face shell of the concrete masonry by dumping from a wheelbarrow. The grout in the keyways between the slabs will be dumped from a wheelbarrow and then spread with long-handled tools, allowing the worker to stand erect facing toward the unprotected edge and back from any work deck edge.

  - ○ Whenever possible, the designated erectors will approach the incoming member at the leading edge only after it is below waist height so that the member itself provides protection against falls.

  - ○ Except for the situations described below, when the arriving floor or roof member is within 2" to 3" of its final position, the designated erectors can then proceed to their position of erection at each end of the member under the control of the safety monitor. Crane hooks will be unhooked from double tee members by designated erectors under the direction and supervision of the safety monitor.

- ○ Designated erectors, while waiting for the next floor or roof member, will be constantly under the control of the safety monitor for fall protection and are directed to stay a minimum of six (6) feet from the edge. In the event a designated erector must move from one end of a member, which has just been placed at the leading edge, he must first move away from the leading edge a minimum of six (6) feet and then progress to the other end while maintaining the minimum distance of six (6) feet at all times.

- ○ Erection of double tees, where conditions require bearing of one end into a closed pocket and the other end on a beam ledge, restricting the tee legs from going directly into the pockets, requires special considerations. The tee legs that are to bear in the closed pocket must hang lower than those at the beam bearing. The double tee will be "two-lined" in order to elevate one end higher than the other to allow for the low end to be ducked into the closed pocket using the following procedure.

- ○ The double tee will be rigged with a standard four-way spreader off of the main load line. An additional choker will be attached to the married point of the two-legged spreader at the end of the tee that is to be elevated. The double tee will be hoisted with the main load line and swung into a position as close as possible to the tee's final bearing elevation. When the tee is in this position and stabilized, the whip line load block will be lowered to just above the tee deck. At this time, two erectors will walk out on the suspended tee deck at midspan of the tee member and pull the load block to the end of the tee to be elevated and attach the additional choker to the load block. The possibility of entanglement with the crane lines and other obstacles during this two-lining process while raising and lowering the crane block on that second line could be hazardous to an encumbered employee. Therefore, the designated erectors will not tie off during any part of this process. While the designated erectors are on the double tee, the safety monitoring system will be used. After attaching the choker, the two erectors then step back on the previously erected tee deck and signal the crane operator to hoist the load with the whip line to the elevation that will allow for enough clearance to let the low-end tee legs slide into the pockets when the main load line is lowered. The erector, who is handling the lowered end of the tee at the closed pocket bearing, will step out on the suspended tee. An erection bar will then be placed between the end of the tee leg and the inside face of the pocketed spandrel member. The tee is barred away from the pocketed member to reduce the friction and lateral force against the pocketed member. As the tee is being lowered, the other erector remains on the tee which was previously erected to handle the other end. At this point the tee is slowly lowered by the crane to a point where the tee legs can freely slide into the pockets. The erector working the lowered end of the tee must keep pressure on the bar between the tee and the face of the pocketed spandrel member to very gradually let the tee legs slide into the pocket to its proper bearing dimension. The tee is then slowly lowered into its final erected position.

- ○ The designated erector should be allowed onto the suspended double tee; otherwise, there is no control over the horizontal movement of the double tee and this movement could knock the spandrel off of its bearing or the column out of plumb. The control necessary to prevent hitting the spandrel can only be done safely from the top of the double tee being erected.

- ○ Load-bearing Wall Panels: The erection of the load-bearing wall panels on the elevated decks requires the use of a safety monitor and a controlled access zone that is a minimum of 25' and a maximum of one-half the length of the wall panels away from the unprotected edge so that designated erectors can move freely and unencumbered when receiving the panels. Bracing, if required for stability, will be installed by ladder. After the braces are secured, the crane will be disconnected from the wall by using a ladder. The wall to wall connections will also be performed from a ladder.

       ○ Non-Load-bearing Panels (Cladding): The locating of survey lines, panel layout, and other installation prerequisites (preceding, etc.) for non-load-bearing panels (cladding) will not commence until floor perimeter and floor openings have been protected. In some areas, it is necessary because of panel configuration to remove the perimeter protection as the cladding is being installed. Removal of perimeter protection will be performed on a bay-to-bay basis, just ahead of cladding erection to minimize temporarily unprotected floor edges. Those workers within 6' of the edge receiving and positioning the cladding when the perimeter protection is removed shall be tied off.

- Detailing

       ○ Employees exposed to falls of six (6) feet or more to lower levels who are not actively engaged in leading edge work or connecting activity, such as welding, bolting, cutting, bracing, guying, patching, painting, or other operations, and who are working less than six (6) feet from an unprotected edge will be tied off at all times or guardrails will be installed. Employees engaged in these activities but who are more than six (6) feet from an unprotected edge as defined by the control zone lines do not require fall protection, but a warning line or control lines must be erected to remind employees they are approaching an area where fall protection is required.

IV. Conventional Fall Protection Considered for the Point of Erection or Leading Edge Erection Operations

A. Personal Fall Arrest Systems

In this particular erection sequence and procedure, personal fall arrest systems requiring body belt/harness systems, lifelines, and lanyards will not reduce possible hazards to workers and will create offsetting hazards during their usage at the leading edge of precast/pre-stressed concrete construction.

Leading edge erection and initial connections are conducted by employees who are specifically trained to do this type of work and are trained to recognize the fall hazards. The nature of such work normally exposes the employee to the fall hazard for a short period of time and installation of fall protection systems for a short duration is not feasible because it exposes the installers of the system to the same fall hazard, but for a longer period of time.

1. It is necessary that the employee be able to move freely without encumbrance in order to guide the sections of precast concrete into their final position without having lifelines attached that will restrict the employee's ability to move about at the point of erection.

2. A typical procedure requires two or more workers to maneuver around each other as a concrete member is positioned to fit into the structure. If they are each attached to a lifeline, part of their attention must be diverted from their main task of positioning a member weighing several tons to the task of avoiding entanglements of their lifelines or avoiding tripping over lanyards. Therefore, if these workers are attached to lanyards, more fall potential would result than from not using such a device.

In this specific erection sequence and procedure, retractable lifelines do not solve the problem of two workers becoming tangled. In fact, such a tangle could prevent the lifeline from retracting as the worker moved, thus potentially exposing the worker to a fall greater than 6'. Also, a worker crossing over the lifeline of another worker can create a hazard because the movement of one person can unbalance the other. In the event of a fall by one person, there is a likelihood that the other person will be caused to fall as well. In addition, if contamination such as grout (during hollow-core grouting) enters the retractable housing, it can cause excessive wear and damage to the device and could clog the retracting mechanism as the lanyard is dragged across the deck. Obstructing the cable orifice can defeat the device's shock-absorbing function, produce cable slack and damage, and adversely affect cable extraction and retraction.

3. Employees tied to a lifeline can be trapped and crushed by moving structural members if the employee becomes restrained by the lanyard or retractable lifeline and cannot get out of the path of the moving load. The sudden movement of a precast concrete member being raised by a crane can be caused by a number of factors. When this happens, a connector may immediately have to move a considerable distance to avoid injury. If a tied-off body belt/harness is being used, the connector could be trapped. Therefore, there is a greater risk of injury if the connector is tied to the structure for this specific erection sequence and procedure.

   When necessary to move away from a retractable device, the worker cannot move at a rate greater than the device locking speed, typically 3.5 to 4.5 ft/sec. When moving toward the device it is necessary to move at a rate that does not permit cable slack to build up. This slack may cause cable retraction acceleration and cause a worker to lose balance by applying a higher-than-normal jerking force on the body when the cable suddenly becomes taut after building up momentum. This slack can also cause damage to the internal spring-loaded drum, uneven coiling of cable on the drum, and possible cable damage.

   The factors causing sudden movements for this location include:

   (a) Cranes

      (1) Operator error.

      (2) Site conditions (soft or unstable ground).

      (3) Mechanical failure.

      (4) Structural failure.

      (5) Rigging failure.

      (6) Crane signal/radio communication failure.

   (b) Weather Conditions

      (1) Wind (strong wind/sudden gusting)—particularly a problem with the large surface areas of precast concrete members.

      (2) Snow/rain (visibility).

      (3) Fog (visibility).

      (4) Cold—causing slowed reactions or mechanical problems.

   (c) Structure/Product Conditions

      (1) Lifting eye failure.

      (2) Bearing failure or slippage.

      (3) Structure shifting.

      (4) Bracing failure.

      (5) Product failure.

   (d) Human Error

      (1) Incorrect tag line procedure.

      (2) Tag line hang-up.

      (3) Incorrect or misunderstood crane signals.

      (4) Misjudged elevation of member.

      (5) Misjudged speed of member.

      (6) Misjudged angle of member.

4. Anchorages or special attachment points could be cast into the precast concrete members if sufficient preplanning and consideration of erectors' positions are done before the members are cast. Any hole or other attachment must be approved by the engineer who designed the member. It is possible that some design restrictions will not allow a member to be weakened by an additional hole; however, it is anticipated that such situations would be the exception, not the rule. Attachment points, other than on the deck surface, will require removal and/or patching. In order to remove and/or patch these points, the employees are exposed to an additional fall hazard at an unprotected perimeter. The fact that attachment points could be available anywhere on the structure does not eliminate the hazards of using these points for tying off, as discussed above. A logical point for tying off on double tees would be using the lifting loops, except that they must be cut off to eliminate a tripping hazard at an appropriate time.

5. Providing attachment at a point above the walking/working surface would also create fall exposures for employees installing their devices. Final positioning of a precast concrete member requires it to be moved in such a way that it must pass through the area that would be occupied by the lifeline and the lanyards attached to the point above. Resulting entanglements of lifelines and lanyards on a moving member could pull employees from the work surface. Also, the structure is being created and, in most cases, there is no structure above the members being placed.

   (a) Temporary structural supports, installed to provide attaching points for lifelines limit the space which is essential for orderly positioning, alignment, and placement of the precast concrete members. To keep the lanyards a reasonable and manageable length, lifeline supports would necessarily need to be in proximity to the positioning process. A sudden shift of the precast concrete member being positioned because of wind pressure or crane movement could make it strike the temporary supporting structure, moving it suddenly and causing tied off employees to fall.

   (b) The time in manhours that would be expended in placing and maintaining temporary structural supports for lifeline attaching points could exceed the expended manhours involved in placing the precast concrete members. No protection could be provided for the employees erecting the temporary structural supports, and these supports would have to be moved for each successive step in the construction process, thus greatly increasing the employees' exposure to the fall hazard.

   (c) The use of a cable strung horizontally between two columns to provide tie-off lines for erecting or walking a beam for connecting work is not feasible and creates a greater hazard on this multistory building for the following reasons:

       (1) If a connector is to use such a line, it must be installed between the two columns. To perform this installation requires an erector to have more fall exposure time attaching the cable to the columns than would be spent to make the beam-to-column connection itself.

       (2) If such a line is to be installed so that an erector can walk along a beam, it must be overhead or below him. For example, if a connector must walk along a 24"-wide beam, the presence of a line next to the connector at waist level, attached directly to the columns, would prevent connectors from centering their weight over the beam and balancing themselves. Installing the line above the connector might be possible on the first level of a two-story column; however, the column may extend only a few feet above the floor level at the second level or be flush with the floor level. Attaching the line to the side of the beam could be a solution; however, it would require the connector to attach the lanyard below foot level, which would most likely extend a fall farther than 6'.

       (3) When lines are strung over every beam, it becomes more and more difficult for the crane operator to lower a precast concrete member into position without the member becoming fouled. Should the member become entangled, it could easily dislodge the line from a column. If a worker is tied to it at the time, a fall could be caused.

6. The ANSI A10.14-1991, American National Standard for Construction and Demolition Operations—Requirements for Safety Belts, Harnesses, Lanyards, and Lifelines for Construction and Demolition Use, states that the anchor point of a lanyard or deceleration device should, if possible, be located above the wearer's belt or harness attachment. ANSI A10.14 also states that a suitable anchorage point is one that is located as high as possible to prevent contact with an obstruction below should the worker fall. Most manufacturers also warn in the user's handbook that the safety block/retractable lifeline must be positioned above the D-ring (above the work space of the intended user), and OSHA recommends that fall arrest and restraint equipment be used in accordance with the manufacturer's instructions.

Attachment of a retractable device to a horizontal cable near floor level or using the inserts in the floor or roof members may result in increased free fall due to the dorsal D-ring of the full-body harness riding higher than the attachment point of the snaphook to the cable or insert (e.g., 6-foot-tall worker with a dorsal D-ring at 5 feet above the floor or surface reduces the working length to only 1 foot by placing the anchorage 5 feet away from the fall hazard). In addition, impact loads may exceed maximum fall arrest forces (MAF) because the fall arrest D-ring would be 4 to 5 feet higher than the safety block/retractable lifeline anchored to the walking/working surface and the potential for swing hazards is increased. Manufacturers also require that workers not work at a level where the point of snaphook attachment to the body harness is above the device because this will increase the free-fall distance and the deceleration distance and will cause higher forces on the body in the event of an accidental fall.

Manufacturers recommend an anchorage for the retractable lifeline that is immovably fixed in space and is independent of the user's support systems. A moveable anchorage is one that can be moved around (such as equipment or wheeled vehicles) or which can deflect substantially under shock loading (such as a horizontal cable or very flexible beam). In the case of a very flexible anchorage, a shock load applied to the anchorage during fall arrest can cause oscillation of the flexible anchorage such that the retractable brake mechanism may undergo one or more cycles of locking/unlocking/locking (ratchet effect) until the anchorage deflection is dampened. Therefore, use of a moveable anchorage involves critical engineering and safety factors and should only be considered after fixed anchorage has been determined to be not feasible.

Horizontal cables used as an anchorage present an additional hazard due to amplification of the horizontal component of maximum arrest force (of a fall) transmitted to the points where the horizontal cable is attached to the structure. This amplification is due to the angle of sag of a horizontal cable and is most severe for small angles of sag. For a cable sag angle of 2 degrees, the horizontal force on the points of cable attachment can be amplified by a factor of 15.

It is also necessary to install the retractable device vertically overhead to minimize swing falls. If an object is in the worker's swing path (or that of the cable), hazardous situations exist: (1) due to the swing, the horizontal speed of the user may be high enough to cause injury when an obstacle in the swing fall path is struck by either the user or the cable; (2) the total vertical fall distance of the user may be much greater than if the user had fallen only vertically without a swing fall path.

With retractable lines, overconfidence may cause the worker to engage in inappropriate behavior, such as approaching the perimeter of a floor or roof at a distance appreciably greater than the shortest distance between the anchorage point and the leading edge. Though the retractable lifeline may arrest a worker's fall before he or she has fallen a few feet, the lifeline may drag along the edge of the floor or beam and swing the worker like a pendulum until the line has moved to a position where the distance between the anchorage point and floor edge is the shortest distance between those two points. Accompanying this pendulum swing is a lowering of the worker, with the attendant danger that he or she may violently impact the floor or some obstruction below. The risk of a cable breaking is increased if a lifeline is dragged sideways across the rough surface or edge of a concrete member at the same moment that the lifeline is being subjected to a maximum impact loading during a fall. The typical 3/16" cable in a retractable lifeline has a breaking strength of from 3000 to 3700 lbs.

7. The competent person, who can take into account the specialized operations being performed on this project, should determine when and where a designated erector cannot use a personal fall arrest system.

B. Safety Net Systems

The nature of this particular precast concrete erection worksite precludes the safe use of safety nets where point of erection or leading edge work must take place.

1. To install safety nets in the interior high bay of the single-story portion of the building poses rigging attachment problems. Structural members do not exist to which supporting devices for nets can be attached in the area where protection is required. As the erection operation advances, the location of point of erection or leading edge work changes constantly as each member is attached to the structure. Due to this constant change it is not feasible to set net sections and build separate structures to support the nets.

2. The nature of the erection process for the precast concrete members is such that an installed net would protect workers as they position and secure only one structural member. After each member is stabilized, the net would have to be moved to a new location (this could mean a move of 8' to 10' or the possibility of a move to a different level or area of the structure) to protect workers placing the next piece in the construction sequence. The result would be the installation and dismantling of safety nets repeatedly throughout the normal workday. As the time necessary to install a net, test, and remove it is significantly greater than the time necessary to position and secure a precast concrete member, the exposure time for the worker installing the safety net would be far longer than for the workers whom the net is intended to protect. The time exposure repeats itself each time the nets and supporting hardware must be moved laterally or upward to provide protection at the point of erection or leading edge.

3. Strict interpretation of 1926.502(c) requires that operations shall not be undertaken until the net is in place and has been tested. With the point of erection constantly changing, the time necessary to install and test a safety net significantly exceeds the time necessary to position and secure the concrete member.

4. Use of safety nets on exposed perimeter wall openings and open-sided floors causes attachment points to be left in architectural concrete that must be patched and filled with matching material after the net-supporting hardware is removed. In order to patch these openings, additional numbers of employees must be suspended by swing stages, boatswain chairs, or other devices, thereby increasing the amount of fall exposure time to employees.

5. Installed safety nets pose an additional hazard at the perimeter of the erected structure where limited space is available in which members can be turned after being lifted from the ground by the crane. There would be a high probability that the member being lifted could become entangled in net hardware, cables, etc.

6. The use of safety nets where structural wall panels are being erected would prevent movement of panels to the point of installation. To be effective, nets would necessarily have to provide protection across the area where structural supporting wall panels would be set and plumbed before roof units could be placed.

7. Use of a tower crane for the erection of the high-rise portion of the structure poses a particular hazard in that the crane operator cannot see or judge the proximity of the load in relation to the structure or nets. If the signaler is looking through nets and supporting structural devices while giving instructions to the crane operator, it is not possible to judge precise relationships between the load and the structure itself or to nets and supporting structural devices. This could cause the load to become entangled in the net or hit the structure, causing potential damage.

C. Guardrail Systems

On this particular worksite, guardrails, barricades, ropes, cables, or other perimeter guarding devices or methods on the erection floor will pose problems to safe erection procedures. Typically, a floor or roof is erected by placing 4'- to 10'-wide structural members next to one another and welding or grouting them together. The perimeter of a floor and roof changes each time a new member is placed into position. It is unreasonable and virtually impossible to erect guardrails and toe boards at the ever-changing leading edge of a floor or roof.

1. To position a member safely it is necessary to remove all obstructions extending above the floor level near the point of erection. Such a procedure allows workers to swing a new member across the erected surface as necessary to position it properly without worrying about knocking material off of this surface.

   Hollow-core slab erection on the masonry wall requires installation of the perimeter protection where the masonry wall has to be constructed. This means the guardrail is installed and then subsequently removed to continue the masonry construction. The erector will be exposed to a fall hazard for a longer period of time while installing and removing perimeter protection than while erecting the slabs.

   In hollow-core work, as in other precast concrete erection, others are not typically on the work deck until the precast concrete erection is complete. The deck is not complete until the leveling, aligning, and grouting of the joints is done. It is normal practice to keep others off the deck until at least the next day after the installation is complete to allow the grout to harden.

2. There is no permanent boundary until all structural members have been placed in the floor or roof. At the leading edge, workers are operating at the temporary edge of the structure as they work to position the next member in the sequence. Compliance with the standard would require a guardrail and toe board be installed along this edge. However, the presence of such a device would prevent a new member from being swung over the erected surface low enough to allow workers to control it safely during the positioning process. Further, these employees would have to work through the guardrail to align the new member and connect it to the structure. The guardrail would not protect an employee who must lean through it to do the necessary work; rather, it would hinder the employee to such a degree that a greater hazard is created than if the guardrail were absent.

3. Guardrail requirements pose a hazard at the leading edge of installed floor or roof sections by creating the possibility of employees being caught between guardrails and suspended loads. The lack of a clear work area in which to guide the suspended load into position for placement and welding of members into the existing structure creates still further hazards.

4. Where erection processes require precast concrete stairways or openings to be installed as an integral part of the overall erection process, it must also be recognized that guardrails or handrails must not project above the surface of the erection floor. Such guardrails should be terminated at the level of the erection floor to avoid placing hazardous obstacles in the path of a member being positioned.

V. Other Fall Protection Measures Considered for this Job

The following is a list and explanation of other fall protection measures available and an explanation of limitations for use on this particular jobsite. If during the course of erecting the building the employee sees an area that could be erected more safely by the use of these fall protection measures, the foreman should be notified.

A. Scaffolds are not used because:

1. The leading edge of the building is constantly changing and the scaffolding would have to be moved at very frequent intervals. Employees erecting and dismantling the scaffolding would be exposed to fall hazards for a greater length of time than they would by merely erecting the precast concrete member.

2. A scaffold tower could interfere with the safe swinging of a load by the crane.

3. Power lines, terrain, and site do not allow for the safe use of scaffolding.

B. Vehicle-mounted platforms are not used because:

1. A vehicle-mounted platform will not reach areas on the deck that are erected over other levels.

2. The leading edge of the building is usually over a lower level of the building and this lower level will not support the weight of a vehicle-mounted platform.

3. A vehicle-mounted platform could interfere with the safe swinging of a load by the crane, either by the crane swinging the load over or into the equipment.

4. Power lines and surrounding site work do not allow for the safe use of a vehicle-mounted platform.

C. Crane-suspended personnel platforms are not used because:

1. A second crane close enough to suspend any employee in the working and erecting area could interfere with the safe swinging of a load by the crane hoisting the product to be erected.

2. Power lines and surrounding site work do not allow for the safe use of a second crane on the job.

VI. Enforcement

Constant awareness of and respect for fall hazards and compliance with all safety rules are considered conditions of employment. The jobsite Superintendent, as well as individuals in the Safety and Personnel Department, reserves the right to issue disciplinary warnings to employees, up to and including termination, for failure to follow the guidelines of this program.

VII. Accident Investigations

All accidents that result in injury to workers, regardless of their nature, shall be investigated and reported. It is an integral part of any safety program that documentation take place as soon as possible so that the cause and means of prevention can be identified to prevent a reoccurrence.

In the event that an employee falls or there is some other, related serious incident occurring, this plan shall be reviewed to determine if additional practices, procedures, or training need to be implemented to prevent similar types of falls or incidents from occurring.

VIII. Changes to Plan

Any changes to the plan will be approved by (name of the qualified person). This plan shall be reviewed by a qualified person as the job progresses to determine if additional practices, procedures, or training need to be implemented by the competent person to improve or provide additional fall protection. Workers shall be notified and trained, if necessary, in the new procedures. A copy of this plan and all approved changes shall be maintained at the jobsite.

SAMPLE FALL PROTECTION PLAN FOR RESIDENTIAL CONSTRUCTION

OSHA's sample fall protection plan for residential construction projects can be found at http://www.osha.gov, as well as dewalt.com/guides.

SAMPLE FALL PROTECTION PLAN FOR RESIDENTIAL CONSTRUCTION

(INSERT COMPANY NAME)

This Fall Protection Plan is specific for the following project:

Location of Job:

Date Plan Prepared or Modified:

Plan Prepared by:

Plan Approved by:

Plan Supervised by:

The following Fall Protection Plan is a sample program prepared for the prevention of injuries associated with falls. A Fall Protection Plan must be developed and evaluated on a site-by-site basis. It is recommended that builders discuss the written Fall Protection Plan with their OSHA Area Office prior to going on a jobsite.

I. Statement of Company Policy

(Your company name here) is dedicated to the protection of its employees from on-the-job injuries. All employees of (your company name here) have the responsibility to work safely on the job. The purpose of the plan is to supplement our existing safety and health program and to ensure that every employee who works for (your company name here) recognizes workplace fall hazards and takes the appropriate measures to address those hazards.

This Fall Protection Plan addresses the use of conventional fall protection at a number of areas on the project, as well as identifies specific activities that require nonconventional means of fall protection. During the construction of residential buildings under 48 feet in height, it is sometimes infeasible or it creates a greater hazard to use conventional fall protection systems at specific areas or for specific tasks. The areas or tasks may include, but are not limited to:

a. Setting and bracing of roof trusses and rafters;

b. Installation of floor sheathing and joists;

c. Roof sheathing operations; and

d. Erecting exterior walls.

In these cases, conventional fall protection systems may not be the safest choice for builders. This plan is designed to enable employers and employees to recognize the fall hazards associated with this job and to establish the safest procedures that are to be followed in order to prevent falls to lower levels or through holes and openings in walking/working surfaces.

Each employee will be trained in these procedures and will strictly adhere to them except when doing so would expose the employee to a greater hazard. If, in the employee's opinion, this is the case, the employee is to notify the competent person of their concern and have the concern addressed before proceeding.

It is the responsibility of (name of competent person) to implement this Fall Protection Plan. Continual observational safety checks of work operations and the enforcement of the safety policy and procedures shall be regularly enforced. The crew supervisor or foreman (insert name) is responsible for correcting any unsafe practices or conditions immediately.

It is the responsibility of the employer to ensure that all employees understand and adhere to the procedures of this plan and to follow the instructions of the crew supervisor. It is also the responsibility of the employee to bring to management's attention any unsafe or hazardous conditions or practices that may cause injury to either themselves or any other employees. Any changes to the Fall Protection Plan must be approved by (name of qualified person).

II. Fall Protection Systems to Be Used on This Job

Installation of roof trusses/rafters, exterior wall erection, roof sheathing, floor sheathing, and joist/truss activities will be conducted by employees who are specifically trained to do this type of work and are trained to recognize the fall hazards. The nature of such work normally exposes the employee to the fall hazard for a short period of time. This plan details how (your company name here) will minimize these hazards.

Controlled Access Zones

When using the plan to implement the fall protection options available, workers must be protected through limited access to high hazard locations. Before any nonconventional fall protection systems are used as part of the work plan, a controlled access zone (CAZ) shall be clearly defined by the competent person as an area where a recognized hazard exists. The demarcation of the CAZ shall be communicated by the competent person in a recognized manner, either through signs, wires, tapes, ropes, or chains.

(Your company name here) shall take the following steps to ensure that the CAZ is clearly marked or controlled by the competent person:

- All access to the CAZ must be restricted to authorized entrants;

- All workers who are permitted in the CAZ shall be listed in the appropriate sections of the plan (or be visibly identifiable by the competent person) prior to implementation;

- The competent person shall ensure that all protective elements of the CAZ be implemented prior to the beginning of work.

Installation Procedures for Roof Truss and Rafter Erection

During the erection and bracing of roof trusses/rafters, conventional fall protection may present a greater hazard to workers. On this job, safety nets, guardrails and personal fall arrest systems will not provide adequate fall protection because the nets will cause the walls to collapse, while there are no suitable attachment or anchorage points for guardrails or personal fall arrest systems.

On this job, requiring workers to use a ladder for the entire installation process will cause a greater hazard because the worker must stand on the ladder with his back or side to the front of the ladder. While erecting the truss or rafter the worker will need both hands to maneuver the truss and therefore cannot hold onto the ladder. In addition, ladders cannot be adequately protected from movement while trusses are being maneuvered into place. Many workers may experience additional fatigue because of the increase in overhead work with heavy materials, which can also lead to a greater hazard.

Exterior scaffolds cannot be utilized on this job because the ground, after recent backfilling, cannot support the scaffolding. In most cases, the erection and dismantling of the scaffold would expose workers to a greater fall hazard than erection of the trusses/rafters.

On all walls of 8 feet or less, workers will install interior scaffolds along the interior wall below the location where the trusses/rafters will be erected. "Sawhorse" scaffolds constructed of 46 inch sawhorses and 2×10 planks will often allow workers to be elevated high enough to allow for the erection of trusses and rafters without working on the top plate of the wall.

In structures that have walls higher than 8 feet and where the use of scaffolds and ladders would create a greater hazard, safe working procedures will be utilized when working on the top plate and will be monitored by the crew supervisor. During all stages of truss/rafter erection the stability of the trusses/rafters will be ensured at all times.

(Your company name here) shall take the following steps to protect workers who are exposed to fall hazards while working from the top plate installing trusses/rafters:

- Only the following trained workers will be allowed to work on the top plate during roof truss or rafter installation:
  -
  -
  -
  -
  -
  -

- Workers shall have no other duties to perform during truss/rafter erection procedures;

- All trusses/rafters will be adequately braced before any worker can use the truss/rafter as a support;

- Workers will remain on the top plate using the previously stabilized truss/rafter as a support while other trusses/rafters are being erected;

- Workers will leave the area of the secured trusses only when it is necessary to secure another truss/rafter;

- The first two trusses/rafters will be set from ladders leaning on side walls at points where the walls can support the weight of the ladder; and

- A worker will climb onto the interior top plate via a ladder to secure the peaks of the first two trusses/rafters being set.

The workers responsible for detaching trusses from cranes and/or securing trusses at the peaks traditionally are positioned at the peak of the trusses/rafters. There are also situations where workers securing rafters to ridge beams will be positioned on top of the ridge beam.

(Your company name here) shall take the following steps to protect workers who are exposed to fall hazards while securing trusses/rafters at the peak of the trusses/ridge beam:

- Only the following trained workers will be allowed to work at the peak during roof truss or rafter installation:
  -
  -

- ○
- ○
- ○
- ○
- Once truss or rafter installation begins, workers not involved in that activity shall not stand or walk below or adjacent to the roof opening or exterior walls in any area where they could be struck by falling objects;
- Workers shall have no other duties than securing/bracing the trusses/ridge beam;
- Workers positioned at the peaks or in the webs of trusses or on top of the ridge beam shall work from a stable position, either by sitting on a "ridge seat" or other equivalent surface that provides additional stability or by positioning themselves in previously stabilized trusses/rafters and leaning into and reaching through the trusses/rafters;
- Workers shall not remain on or in the peak/ridge any longer than necessary to safely complete the task.

Roof Sheathing Operations

Workers typically install roof sheathing after all trusses/rafters and any permanent truss bracing is in place. Roof structures are unstable until some sheathing is installed, so workers installing roof sheathing cannot be protected from fall hazards by conventional fall protection systems until it is determined that the roofing system can be used as an anchorage point. At that point, employees shall be protected by a personal fall arrest system.

Trusses/rafters are subject to collapse if a worker falls while attached to a single truss with a belt/harness. Nets could also cause collapse, and there is no place to attach guardrails.

All workers will ensure that they have secure footing before they attempt to walk on the sheathing, including cleaning shoes/boots of mud or other slip hazards.

To minimize the time workers must be exposed to a fall hazard, materials will be staged to allow for the quickest installation of sheathing.

(Your company name here) shall take the following steps to protect workers who are exposed to fall hazards while installing roof sheathing:

- Once roof sheathing installation begins, workers not involved in that activity shall not stand or walk below or adjacent to the roof opening or exterior walls in any area where they could be struck by falling objects;
- The competent person shall determine the limits of this area, which shall be clearly communicated to workers prior to placement of the first piece of roof sheathing;
- The competent person may order work on the roof to be suspended for brief periods as necessary to allow other workers to pass through such areas when this would not create a greater hazard;
- Only qualified workers shall install roof sheathing;
- The bottom row of roof sheathing may be installed by workers standing in truss webs;
- After the bottom row of roof sheathing is installed, a slide guard extending the width of the roof shall be securely attached to the roof. Slide guards are to be constructed of no less than nominal 4" height capable of limiting the uncontrolled slide of workers. Workers should install the slide guard while standing in truss webs and leaning over the sheathing;
- Additional rows of roof sheathing may be installed by workers positioned on previously installed rows of sheathing. A slide guard can be used to assist workers in retaining their footing during successive sheathing operations; and

- Additional slide guards shall be securely attached to the roof at intervals not to exceed 13 feet as successive rows of sheathing are installed. For roofs with pitches in excess of 9-in-12, slide guards will be installed at 4-foot intervals.

When wet weather (rain, snow, or sleet) are present, roof sheathing operations shall be suspended unless safe footing can be assured for those workers installing sheathing.

When strong winds (above 40 miles per hour) are present, roof sheathing operations are to be suspended unless wind breakers are erected.

Installation of Floor Joists and Sheathing

During the installation of floor sheathing/joists (leading edge construction), the following steps shall be taken to protect workers:

Only the following trained workers will be allowed to install floor joists or sheathing:

- Employee Name (s):
  - ○
  - ○
  - ○
  - ○
  - ○
  - ○

- Materials for the operations shall be conveniently staged to allow for easy access to workers;
- The first floor joists or trusses will be rolled into position and secured either from the ground, ladders or sawhorse scaffolds;
- Each successive floor joist or truss will be rolled into place and secured from a platform created from a sheet of plywood laid over the previously secured floor joists or trusses;
- Except for the first row of sheathing, which will be installed from ladders or the ground, workers shall work from the established deck; and
- Any workers not assisting in the leading edge construction while leading edges still exist (e.g., cutting the decking for the installers) shall not be permitted within 6 feet of the leading edge under construction.

Erection of Exterior Walls

During the construction and erection of exterior walls, employers shall take the following steps to protect workers:

- Only the following trained workers will be allowed to erect exterior walls:
  - ○
  - ○
  - ○
  - ○
  - ○
  - ○

- A painted line 6 feet from the perimeter will be clearly marked prior to any wall erection activities to warn of the approaching unprotected edge;

- Materials for operations shall be conveniently staged to minimize fall hazards; and
- Workers constructing exterior walls shall complete as much cutting of materials and other preparation as possible away from the edge of the deck.

III. Enforcement

Constant awareness of and respect for fall hazards, and compliance with all safety rules are considered conditions of employment. The crew supervisor or foreman, as well as individuals in the Safety and Personnel Department, reserve the right to issue disciplinary warnings to employees, up to and including termination, for failure to follow the guidelines of this program.

IV. Accident Investigations

All accidents that result in injury to workers, regardless of their nature, shall be investigated and reported. It is an integral part of any safety program that documentation take place as soon as possible so that the cause and means of prevention can be identified to prevent a reoccurrence.

In the event that an employee falls or there is some other related serious incident occurring, this plan shall be reviewed to determine if additional practices, procedures, or training needs to be implemented to prevent similar types of falls or incidents from occurring.

V. Changes to Plan

Any changes to the plan will be approved by (name of the qualified person). This plan shall be reviewed by a qualified person as the job progresses to determine if additional practices, procedures, or training needs to be implemented by the competent person to improve or provide additional fall protection. Workers shall be notified and trained, if necessary, in the new procedures. A copy of this plan and all approved changes shall be maintained at the jobsite.

FIELD METHOD FOR IDENTIFYING SOIL TEXTURE

This OSHA form may be found at http://www.osha.gov or dewalt.com/guides.

Soil Texture	Visual Detection of Particle Size and General Appearance of the Soil	Squeezed in Hand and Pressure Released		Soil Ribboned Between Thumb and Finger When Moist
		When Air Dry	When Moist	
Sand	Soil has a granular appearance in which the individual grain sizes can be detected. It is free-flowing when in a dry condition.	Will not form a cast and will fall apart when pressure is released.	Forms a cast that will crumble when lightly touched.	Cannot be ribboned.
Sandy Loam	Essentially a granular soil with sufficient silt and clay to make it somewhat coherent. Sand characteristics predominate.	Forms a cast that readily falls apart when lightly touched.	Forms a cast that will bear careful handling without breaking.	Cannot be ribboned.
Loam	A uniform mixture of sand, silt, and clay. Grading of sand fraction quite uniform from coarse to fine. It is mellow, has somewhat gritty feel, yet is fairly smooth and slightly plastic.	Forms a cast that will bear careful handling without breaking.	Forms a cast that can be handled freely without breaking.	Cannot be ribboned.

Silt Loam	Contains a moderate amount of the finer grades of sand and only a small amount of clay. Over half of the particles are silt. When dry it may appear quite cloddy, which readily can be broken and pulverized to a powder.	Forms a cast that can be freely handled. Pulverized, it has a soft, flourlike feel.	Forms a cast that can be freely handled. When wet, soil runs together and puddles.	It will not ribbon, but it has a broken appearance, feels smooth, and may be slightly plastic.
Silt	Contains over 80% of silt particles with very little fine sand and clay. When dry, it may be cloddy; readily pulverizes to powder with a soft, flourlike feel.	Forms a cast that can be handled without breaking.	Forms a cast that can freely be handled. When wet, it readily puddles.	It has a tendency to ribbon with a broken appearance; feels smooth.
Clay Loam	Fine-textured soil breaks into very hard lumps when dry. Contains more clay than silt loam. Resembles clay in a dry condition; identification is made on physical behavior of moist soil.	Forms a cast that can be freely handled without breaking.	Forms a cast that can be handled freely without breaking. It can be worked into a dense mass.	Forms a thin ribbon that readily breaks, barely sustaining its own weight.
Clay	Fine-textured soil breaks into very hard lumps when dry. Difficult to pulverize into a soft, flourlike powder when dry. Identification based on cohesive properties of the moist soil.	Forms a cast that can be freely handled without breaking.	Forms a cast that can be handled freely without breaking.	Forms long, thin, flexible ribbons. Can be worked into a dense, compact mass. Considerable plasticity.
Organic Soils	Identification based on the high organic content. Muck consists of thoroughly decomposed organic material with considerable amount of mineral soil finely divided with some fibrous remains. When considerable fibrous material is present, it may be classified as peat. The plant remains or sometimes the woody structure can easily be recognized. Soil color ranges from brown to black. They occur in lowlands, in swamps, or swales. They have high shrinkage upon drying.			

GUIDELINES FOR ESTABLISHING A SITE-SPECIFIC ERECTION PLAN

The following was taken directly from OSHA regulations. The full guidelines can be found at http://www.osha.gov or dewalt.com/guides

GUIDELINES FOR ESTABLISHING THE COMPONENTS OF A SITE-SPECIFIC ERECTION PLAN: NON-MANDATORY GUIDELINES FOR COMPLYING WITH § 1926.752(E).

(a) **General.** This appendix serves as a guideline to assist employers who elect to develop a site-specific erection plan in accordance with § 1926.752(e) with alternate means and methods to provide employee protection in accordance with § 1926.752(e), § 1926.753(c)(5), § 1926.757(a)(4), and § 1926.757(e)(4).

(b) **Development of a site-specific erection plan.** Preconstruction conference(s) and site inspection(s) are held between the erector and the controlling contractor, and others such as the project engineer and fabricator before the start of steel erection. The purpose of such conference(s) is to develop and review the site-specific erection plan that will meet the requirements of this section.

(c) **Components of a site-specific erection plan.** In developing a site-specific erection plan, a steel erector considers the following elements:

    (1) The sequence of erection activity, developed in coordination with the controlling contractor, that includes the following:

        (i) Material deliveries;

        (ii) Material staging and storage; and

        (iii) Coordination with other trades and construction activities.

    (2) A description of the crane and derrick selection and placement procedures, including the following:

        (i) Site preparation;

        (ii) Path for overhead loads; and

        (iii) Critical lifts, including rigging supplies and equipment.

    (3) A description of steel erection activities and procedures, including the following:

        (i) Stability considerations requiring temporary bracing and guying;

        (ii) Erection bridging terminus point;

        (iii) Anchor rod (anchor bolt) notifications regarding repair, replacement, and modifications;

    (iv)   Columns and beams (including joists and purlins);

    (v)   Connections;

    (vi)   Decking; and

    (vii)   Ornamental and miscellaneous iron.

(4)   A description of the fall protection procedures that will be used to comply with § 1926.760.

(5)   A description of the procedures that will be used to comply with § 1926.759.

(6)   A description of the special procedures required for hazardous non-routine tasks.

(7)   A certification for each employee who has received training for performing steel erection operations as required by § 1926.761.

(8)   A list of the qualified and competent persons.

(9)   A description of the procedures that will be utilized in the event of rescue or emergency response.

(d)   **Other plan information.** The plan:

    (1) Includes the identification of the site and project; and

    (2) Is signed and dated by the qualified person(s) responsible for its preparation and modification.

SAMPLE SAFETY INSPECTION CHECKLIST

This sample checklist can help guide you through safety inspections, and is available at dewalt.com/guides.

SAFETY INSPECTION CHECKLIST

This checklist can be used as a starting point. Some construction companies will not need a safety inspection checklist as comprehensive. Feel free to customize this checklist to fit your needs.

Company Name _____

Jobsite Name _____

Jobsite Location _____

Date/Time _____

Superintendent _____

Inspector _____

Current Weather Conditions _____

General	Yes	No	N/A	Date Corrected
Posters and safety signs/warnings?				
Safety meetings held periodically?				
First-aid kit available and adequately stocked?				
Job-related safety training completed?				
Accident reporting procedure established?				
Substance abuse policy in place?				
Injury records being kept?				
Emergency telephone numbers posted?				
Hospital directions posted?				

Housekeeping and Sanitation	Yes	No	N/A	Date Corrected
General housekeeping of work area(s)?				
Containers to handle normal waste?				
Procedures to handle hazardous waste?				
Passageways and walkways clear?				
Adequate lighting?				
Waste containers provided and used?				
Portable toilets adequate and clean?				
Adequate drinking water supply and cups?				
Nails, boards, debris removed and stacked neatly?				
Eyewash facilities available?				
Hazard Communication	**Yes**	**No**	**N/A**	**Date Corrected**
Written program on-site?				
Employees trained?				
MSDSs on file and available?				
Material properly stored and labeled?				
Chemical inventory list on site and available?				
Labels legible?				
Fire Prevention	**Yes**	**No**	**N/A**	**Date Corrected**
Adequate number and type of fire extinguisher(s) available?				
Fire prevention/extinguisher training accomplished?				
Fire extinguisher inspections accomplished (monthly/periodically)?				
Phone number of fire department posted?				
Fire extinguisher(s) provided on appropriate equipment?				
Are flammable liquids in approved containers and correctly labeled?				
Are flammable liquids properly stored?				
Fire alarm available/fire evacuation plan established?				
Fuel supplies protected from accidental impact?				
Fire training given to appropriate personnel?				
Is equipment shut down prior to refueling?				
Is equipment properly grounded to fuel trucks before refueling?				
"No smoking" signs posted and enforced?				
Hydrants clear, access to public thoroughfare open?				
Flammable Liquids/Materials	**Yes**	**No**	**N/A**	**Date Corrected**
Empty containers removed?				
Only approved containers being used?				
Containers stored in approved and appropriate area(s)?				

	Yes	No	N/A	Date Corrected
Outside storage has secondary containment?				
Storage tanks properly grounded, bonded, and pressure relief provided?				
Cylinders stored/secured in upright position?				
Electrical Safety	**Yes**	**No**	**N/A**	**Date Corrected**
Electrical devices have current inspection and coding?				
Electrical equipment properly maintained?				
Equipment properly grounded?				
Assured equipment grounding program established?				
GFCI used and tested where required?				
Electrical dangers posted?				
Are breaker boxes equipped with required covers?				
Are circuits labeled in breaker boxes?				
Personal Protective Equipment	**Yes**	**No**	**N/A**	**Date Corrected**
Employees issued PPE where needed?				
Is PPE being used?				
Employees trained in the use of PPE?				
Adequate fall protection provided?				
Eye protection/Face protection?				
Hearing protection?				
Respirators and masks?				
Head protection?				
Hand and foot protection?				
Hand and Power Tools	**Yes**	**No**	**N/A**	**Date Corrected**
Proper tool used for the job?				
Handles free of cracks and attached to tool properly?				
Inspections and proper maintenance accomplished prior to use?				
Tools neatly stored, safely carried?				
Good housekeeping where tools are used?				
Tools grounded properly or double insulated?				
Guards in place and used correctly?				
Damaged or malfunctioning tools tagged out until repaired or replaced?				
Tools protected from unauthorized use?				
Competent instruction and supervision?				
Cords included in electrical inspection?				
Ladder Safety	**Yes**	**No**	**N/A**	**Date Corrected**
Ladders inspected and in good condition?				
Ladders used properly for type of exposure?				

Ladders secured to prevent slipping, sliding, or falling?				
Do side rails extend 36" above top of landing?				
Are ladders spliced?				
Rungs or cleats not over 12" on center?				
Proper maintenance and storage?				
Are ladders painted?				
Do fixed ladders in excess of 20' have fall protection?				
Are ladders of sufficient strength for the task?				
Scaffolding Safety	**Yes**	**No**	**N/A**	**Date Corrected**
Scaffold erection properly supervised?				
All structural members free from defects and meet safety factor?				
Are all connections secure?				
Are scaffolds erected on solid footing?				
Is scaffold tied to structure?				
Are working areas free of dirt, debris, snow, ice, grease, etc.?				
Are workers protected from falling objects?				
Is scaffold plumb and square, with cross-bracing?				
Are guardrails, midrails, and toe boards in place?				
Fall protection available and in use?				
Excavations and Trenches	**Yes**	**No**	**N/A**	**Date Corrected**
Are holes, trenches, and cuts over 5' deep shored, sloped, or trench boxes used?				
Operation supervised by competent person?				
Spoil banks at least 2' from edges of cut?				
Ladders placed to ensure no greater than 25' of lateral travel by worker?				
Ladder properly secured?				
Are adjacent structures properly shored?				
Is shoring and sheathing correct for soil and depth?				
Are roads and sidewalks supported and protected?				
Excavation barricaded and lighting provided?				
Are equipment ramps adequate?				
Have underground utility installations been identified?				
Registered professional engineer design/approval accomplished?				
Confined-space entry-permit-required plan established?				
Are daily inspections completed by a competent person?				
Cranes and Derricks	**Yes**	**No**	**N/A**	**Date Corrected**
Are annual inspections completed?				
Are operators certified and physical exams current?				

Are daily inspections completed by operators?				
Outriggers used?				
Power lines deactivated, removed, or warning signs posted warning of at least 10' clearance from overhead power lines (voltages 50,000 volts or below)?				
Proper loading for capacity at lifting radius?				
Operation in accordance with manufacturer's instruction?				
Competent person inspecting crane?				
Equipment properly lubricated and maintained?				
Load testing accomplished?				
Signal person qualified and used where needed?				
Riggers qualified?				
Heavy Equipment	**Yes**	**No**	**N/A**	**Date Corrected**
Regular inspection and maintenance?				
Seat belts provided and used in equipment with ROPS?				
Backup alarms working and audible?				
Slow-moving-vehicle emblem attached to rear of equipment operating at less than 25 mph?				
No employees riding equipment without proper seating?				
Lights, brakes, warning signals operative?				
Wheels chocked when necessary?				
Haul roads well maintained and laid out properly?				
Equipment properly secured when not in use?				
Motor Vehicles	**Yes**	**No**	**N/A**	**Date Corrected**
Regular inspection and maintenance?				
Qualified operators?				
Local and state laws observed?				
Brakes, lights, warning devices operative?				
Weight limits and load stress controlled?				
Personnel carried in correct manner?				
All glass in good condition?				
Backup signals provided?				
Fire extinguisher(s) installed?				
Seat belts worn?				
Tie-down straps or chains inspected?				
Are all vehicles checked at the beginning of each shift?				
Welding and Cutting	**Yes**	**No**	**N/A**	**Date Corrected**
Are operators qualified?				
Screens and shields in place?				
Are oxygen and acetylene stored properly?				
Are bottles not in use secured with caps in place?				

Proper eye protection and PPE used?				
Fire extinguisher located near operations?				
"Hot work" permit completed and posted in areas requiring such permit?				
Are valves shut off and regulators backed off each night?				
Flashback arresters placed on hoses (O_2 and fuel gas)?				
Electrical equipment grounded?				
Area inspected for fire hazards?				
Gas lines and power cables protected and in good condition?				
Proper ventilation?				
Steel Erection	**Yes**	**No**	**N/A**	**Date Corrected**
Safety nets or planked floors?				
Hard hats, safety shoes, gloves, and other PPE used?				
Taglines for controlling loads?				
Fire hazards covered and barricaded?				
Floor openings covered and barricaded?				
Hoisting apparatus checked?				
Adequate fall protection?				
Concrete Construction	**Yes**	**No**	**N/A**	**Date Corrected**
Forms properly installed and braced?				
Shoring adequate, plumbed, and cross-braced?				
Proper curing period and procedures used?				
Adequate PPE?				
Caps on rebar?				
Automatic shutoff on power-operated trowels?				
Nails and stripped form material removed from area?				
Masonry Construction	**Yes**	**No**	**N/A**	**Date Corrected**
Proper scaffolding?				
Masonry saws properly equipped, dust protection provided?				
Safe hoisting equipment?				
Are limited-access zones established as required?				
Work Zone Safety	**Yes**	**No**	**N/A**	**Date Corrected**
Competent flagmen properly dressed, trained, and posted?				
Adequate warning signs and markers?				
Equipment not blocking right-of-way?				
Traffic control through construction site?				
Adequate marking and maintenance of detours?				
Dust control used?				
Adequate lighting?				
Are barricades erected with correct directional stripes?				

INDEX

ABOUT THE AUTHOR

Dan Johnson, CSP, is an experienced and integrity-driven Environmental, Health, and Safety (EHS) professional with over 17 years of accomplishments and success on a national level. He is a Certified Safety Professional with the Board of Certified Safety Professionals. Dan is currently the Managing Partner of SFI Compliance Inc. (www.sficompliance.com). He has been a safety director for two national home-building companies and a national utility contractor, and has spent numerous years as a safety consultant specializing in the construction industry. Dan is a skillful designer of EHS programs, and has been responsible for effectively executing these programs within a wide range of industries.

Dan has expertise in changing work environments through behavior-based safety while maintaining productivity. He has extensive knowledge working with multiple jurisdictions to ensure compliance, including: OSHA, DOL, DOT, EPA, USACE, and state and local municipalities.

Dan is also the author of *Jobsite First Aid: A Field Guide for the Construction Industry.*

Dan's experience includes:

- OSHA Authorized Trainer for Construction and General Industry
- CPR/First Aid and AED Train the Trainer
- Fall Protection and Fall Arrest Trainer
- Scaffold Competent Person Trainer
- Hazardous Waste Site Safety (HAZWOPER)
- Confined Space Trainer
- Certified Inspector of Sediment and Erosion Control
- Asbestos Safety Trainer
- Certified Storm Water Inspector
- Forklift Certificate Trainer
- Respiratory Protection Trainer with Fit Testing
- Excavation Competent Person Trainer